ON INFERTILE GROUND

On Infertile Ground

Population Control and Women's Rights in the Era of Climate Change

Jade S. Sasser

NEW YORK UNIVERSITY PRESS
New York

NEW YORK UNIVERSITY PRESS
New York
www.nyupress.org

Library of Congress Cataloging-in-Publication Data
Names: Sasser, Jade S., author.
Title: On infertile ground : population control and women's rights in the era of climate change / Jade Sasser.
Description: New YorK : New York University Press, [2018] |
Includes bibliographical references and index.
Identifiers: LCCN 2018012202 | ISBN 9781479873432 (cl : alk. paper) |
ISBN 9781479899357 (pb : alk. paper)
Subjects: LCSH: Birth control—Environmental aspects. | Population—Environmental aspects. | Climatic changes—Social aspects. | Women's rights. | Feminism.
Classification: LCC HQ766 .S373 2018 | DDC 363.9/6—dc23
LC record available at https://lccn.loc.gov/2018012202

New York University Press books are printed on acid-free paper, and their binding materials are chosen for strength and durability. We strive to use environmentally responsible suppliers and materials to the greatest extent possible in publishing our books.

Manufactured in the United States of America

10 9 8 7 6 5 4 3 2 1

Also available as an ebook

For my parents

CONTENTS

Introduction

Women as Sexual Stewards

One fine day in the spring of 2009, I found myself in a Berkeley, California art gallery for a Sierra Club–sponsored wine and cheese reception. The event was called *Sex and Sustainability*, and it featured presentations by Sierra Club staffers, activist partners, and a local professor, all focused on global population growth, family planning, and the environment. At the time, the global population was well over 6 billion people (we would hit the 7 billion mark two years later), and climate change activists and policymakers had long been frustrated with the U.S. reluctance to join the global policy community in aggressively combating climate change. Meanwhile, the pace of climate change was relentless. Floods, melting glaciers, sea level rise, and threats to wildlife claimed newspaper headlines every week. These weren't just environmental impacts: reported human death tolls in the tens of thousands from intense storms, heat waves, and droughts found their way onto the evening news, illustrating climate change's deadly threat in frightening ways.

However, the reception was upbeat. Barack Obama had been elected president several months prior, ushering in a new era of hope that the U.S. would both increase funding commitments for international family planning, as well as enact binding climate change legislation. The Sierra Club facilitator gave a speech drawing a seamless line of connection between women's fertility, population growth, and environmentalism: "Poor women all over the world are having babies in record numbers, with disastrous impacts on their health, the health of their families, and the environment," she argued. We, the mostly student crowd in the room, had an important part to play in making a difference, by signing up for the Sierra Club's Global Population Environment Program (GPEP) mailing list and connecting to information on various legislative initiatives. "We have to empower women globally, advance access to voluntary family planning, advocate for sexuality and reproductive health education, work to reduce consumption, and support the campaign for international family planning. Oh, and write to Obama!"

Her words highlighted the issue that is central to this book, namely the return of global population to prominence in environmental debates, particularly in the context of climate change. Type "climate change" and "birth control" or "family planning" into your Google search bar, and an endless array of articles proclaiming the climate-solving benefits of contraceptives comes back. Curiously, a number of these articles claim that this solution is new, innovative, or so shrouded in taboo that no one is talking about it. However, this could not be further from the truth. Neo-Malthusians—people who view population growth as the main driver of environmental, social, and economic problems—have been making these arguments for decades, blaming human numbers for everything from deforestation to air pollution, global poverty, civil unrest, international migration, and now climate change. This is a long-enduring narrative that permeates ecological sciences, international development, and everyday conversations about the environment.

What *is* relatively new is the way women's empowerment is being linked to these debates. Population advocates argue that harnessing American foreign aid to provide poor women around the world with universal, voluntary access to contraceptives empowers them to make decisions about their childbearing in ways that affirm their human rights while benefiting the environment by decreasing human numbers. In this schema, fewer people will consume resources and use polluting technologies, relieving pressure on the earth and its atmosphere, which are already being catastrophically stretched to their limits by destructive human activities. While these advocates reject population control because of its historical associations with coercion and human rights abuses, they do maintain that population growth makes environmental and social problems worse, and that their solutions will be easier to achieve if population growth is stabilized.

The distinction, while subtle, is important. Population advocacy arises historically from the deployment of neo-Malthusianism, an expansion of a set of ideas developed in the late 18th century by British cleric and political economist Thomas Robert Malthus. Malthus postulated a theory of the exponential growth of human populations, comparing it to the more limited growth of food production and arguing that human growth far outpaces the earth's capacity to maintain the necessary conditions to sustain human life, leading to inevitable famine and widespread misery. Malthusianism is a political-economic concept couched in the language of biological fundamentalism. Malthus was writing at a time of recent, rapid growth in Britain, primarily among the poor. Debates about state aid to impoverished people were ubiquitous, and Malthus developed his ideas to make a case for why British authorities should remove the state-supported food aid provided to the poor via the British Poor Laws. However, he articulated the problems of population growth and the earth's capacity as functions of nature—describing them as natural law, universal and unchanging.

Twentieth-century neo-Malthusian proponents updated this theory, using it to explain environmental degradation writ large, including everything from toxic air, soil, and water pollution, to deforestation, species extinction, soil erosion, and most recently, climate change. They continue to posit these problems as biological—a natural function and result of human population growth. However, population advocates today reject Malthus's ideas, at least some of them do. One could argue that their position is more closely aligned with what Angus and Butler refer to as "populationism." Like neo-Malthusians, populationists "attribute social and ecological ills to human numbers"[1]; however, they reject coercive population control and demographic targets, and support rights-based solutions, including voluntary access to contraception, access to education, and income-generating opportunities for women and girls worldwide.

Populationism suggests that one can uphold Malthus's and his followers' central claim that there are natural limits to the earth's ability to sustain human life, and that human numbers threaten those limits and must be decreased, while also supporting human rights and international development solutions as the right strategies to slow growth. This populationist perspective is at the heart of population advocacy today. In the words of many advocates I spoke to, family planning programs offer a "win-win" solution for women, population, and the environment. But is it that simple? Are there actual differences between neo-Malthusian and populationist perspectives in terms of how they frame women's relationships to nature, the environment, and reproduction? If so, how do these values shape the kinds of advocacy they develop, the partners they engage, and the outcomes they seek?

Making Sexual Stewards

When population-environment advocates talk about "women" in workshops and other messaging campaigns, they are constructing an idealized model of a woman: a moral agent who manages her fertility and the environment responsibly for the greater good. She is a modern woman who wants two or fewer children, engages in monogamous sexual relationships within the context of marriage, always uses contraceptives consistently and correctly, and makes childbearing decisions in concert with environmental values, including responsible (limited) consumption of energy and natural resources. This is a neoliberal concept: a symbol of the ideal woman framed within the logics of private, individual decision-making and choice, who adopts a modicum of embodied environmental responsibility[2] in the service of global development goals, and who helps advocates weave together narratives of the urgency of simultaneously addressing climate change and empowering women around the world to use contraceptives. She is what I refer to as a *sexual steward*, and she is vital to the future of international family planning policy.

The sexual steward is exemplified in countless reports and articles linking population, climate change, and social justice. Take, for example, a quote from a report produced by the Worldwatch Institute: "Women and children in poverty are among the most vulnerable to the impacts of climate change, despite their disproportionately low contribution to the problem. Removing the obstacles that hold back more than 3 billion potential agents of change—women and girls—is both pragmatic and necessary."[3] The obstacles the report is referring to are reproductive. And the report postulates that removing them will not only improve women's social status, it will potentially solve problems for *the entire world:* "Through slowing growth and other benefits, supporting women's efforts to manage their own lives and improving their status will in turn elevate the well-being of all of the world's population—with Earth's climate representing one aspect of this. And the most effective way to do all this is by making sure, to the extent possible, that women and men everywhere realize their own childbearing intentions, including timing, spacing, and number of children."[4]

This text encapsulates the sexual stewardship idea: "women" are assumed to be fertile, reproducing beings, whose improved status will ideally lead to making responsible family choices—choices that include the *proper* spacing, timing, and number of children that will slow global population growth. Linking these decisions and behaviors to climate change places women's individual reproductive lives in global context. Women's childbearing decisions are thus never individual, never free from the weight of potential environmental catastrophe—and thus never free from a duty to reproduce responsibly. In addition, women in this model are a monolith: sexually active, heterosexual, able to become pregnant and bear children, and free to make their own bodily choices, free of coercion or violence.

As I discovered at the reception described at the start of this chapter, sexual stewards are not just a product of institutional development actors, they also arise from youth activism. Consider the policy statement developed and circulated by youth advocates at the international climate change meetings in Cancun in 2010: "Climate change disproportionately affects women, especially young women, who are often the stewards of their area's natural resources—as they must walk farther to collect water, work harder to produce crops from dry soil, and cope with drought, flooding, [other] natural disasters and disease. At the same time, empowered women can be particularly strong agents for sustainable change in their communities. An effective approach to climate change mitigation and adaptation must support young people's sexual and reproductive health and rights (SRHR), as doing so is essential for adaptation while contributing to reducing the impact of future climate change."[5] Linking environmental problems to women's agency through responsible reproductive management is, apparently, the wave of the future—and the people at the heart of the narrative are young people.

In this book, I argue that sexual stewardship was created to address an international development sector in crisis. Population has been a troubled issue for decades. Decreased funding, histories of coercion, racism, and human rights abuses, and a lack of attention among the broader public have eroded support for what was once a very popular topic of discussion in the U.S. In the mid-twentieth century, population growth was front and center in environmental activism, and had strong ties to the mainstream women's reproductive rights movement. However, as evidence of international and domestic coercion increasingly came to light, coalescing with a conservative political and religious backlash against family planning in the 1980s, population lost its place in the sun in development circles. In the 1990s, transnational feminists organized to offer the international policy community a new way of addressing family planning—emphasizing women's empowerment through voluntary access to contraceptives within broader programs supporting women's SRHR in a human rights framework. This approach is the foundation of sexual stewardship.

Sexual stewardship also rests on instrumental approaches: using a technological solution (contraceptives) to address the complex social, political, and economic drivers of population growth, as well as to build the base of science and activism to support international family planning policy. When women's fertility and reproduction is lifted out of the social contexts of entrenched poverty, gender inequality, growing wealth and income inequality, inadequate access to comprehensive health care, educational services and employment opportunities, and cultural norms favoring large families, it is easy to imagine women as freely acting, autonomous agents whose enduring high fertility is individually driven. In development circles, this model is an attractive way to build a base of new advocates—specifically young environmental activists—as well as to maintain support from those who might otherwise have long abandoned the issue of population.

Given that it rests on such long enduring ideas, is sexual stewardship new? In some ways, it is. Gone is the long-familiar twentieth-century language of population control, through which imminent "global famine," "death," and "destruction" signaled the need for top-down, demographically driven intervention programs. This language has been replaced by a focus on women's rights, justice, and affirmation of the importance of voluntarism to the success of family planning programs. As many population-environment advocates would argue, today's focus on women's reproductive rights and justice evinces a clear break from the so-called dark past of population control—and cements a decades-long shifting of population concerns into the realm of progressive politics.

However, sexual stewardship is in line with a host of neoliberal development strategies focused on population. One such strategy aims to capitalize on the "demographic dividend" produced in a key phase of a nation's demographic transition[6] from high to low population growth. Theorists describe the demographic dividend as an opportunity that arises when, in a transition from high birth and death rates to lower ones, the relative proportion of adults in the labor force is high, compared to their dependents. Their financial resources are more available for investment in the family and the economy, leading to per capita income growth. This is the first dividend. A second potential dividend comes in the form of the increasing longevity of elderly populations, when a concentrated older population accumulates and invests their assets, leading to increased national income. In other words, when there is a higher proportion of working age people relative to the number of dependents, production can increase relative to consumption and GDP per capita may increase.[7] These gains are not automatic: there must be productive workers and consumers, and governments must maximize the opportunities of the population growth window by investing in education, health care, and neoliberal economic policies favoring job markets and future pension programs.[8]

Family planning is central to the demographic dividend strategy in that it slows birth rates and frees workers from responsibilities in the domestic sphere, allowing them to invest more in the formal economy. This approach has gained traction among the international family planning community, and has been included in reports written by the World Health Organization and UN Population Program. However, they promote the demographic dividend as a counterpoint to the "youth bulge" concept, arguing that the youthful demographic dividend must be *managed* in order to be productive and peaceful; otherwise, if not properly supported, it could manifest as a violent bulge of young people, usually men, that threaten global geopolitical security. These concerns mirror mid-twentieth-century Cold War anxieties about the spread of communism among rapidly growing populations in the Global South. At that time, U.S. government leaders, economic planners, and military officials all postulated that slowing population growth in Global South nations through artificial means—Western contraception—was necessary to their economic modernization, as well as to global political stability. The contrasting narratives of demographic dividend and bulge directly rehearse these old discourses, shaping the ways young people access reproductive services. As Hendrixson demonstrates, in the absence of fuller understandings of young people's sexuality and reproduction within the broader context of their lives, dividend-versus-bulge discourses actually constrain young people's access to sexual and reproductive health services, particularly for young men.[9]

In this project, I explore the context in which the sexual stewardship model arose, why it continues to gain in popularity, particularly among college-aged environmental activists, and what possibilities it holds for social justice organizing in the future. Taking the concept of sexual stewardship as my point of departure, my aim is to offer a new way to think about recent population-environment advocacy, and its newest iteration, population-climate advocacy. This advocacy is not comprised of strategies supporting population control, but rather those designed to reframe population interventions as progressive, socially just, and attractive to a new crop of younger environmentalists. In exploring these strategies, I focus on three key trends: recent efforts to build scientific knowledge linking population growth and climate change, making a case for contraceptive intervention from a scientific perspective; the role of private donors behind the scenes; and non-governmental organization (NGO) efforts to enroll youth as population advocates. In particular, I focus my ethnographic observations on a now-defunct training program created by the Sierra Club, known as the Global Population Environment Program. The GPEP was designed to develop a new generation of global population policy advocates through upbeat approaches linking youthful energy, the language of women's empowerment and justice, and policy-relevant science.

This book has three main arguments. First, I argue that science plays an important role in population-climate advocacy, one that functions to legitimize advocates' work and minimize controversy. The more scientific studies that are produced to demonstrate linkages between population growth and climate change, the more effectively the political underpinnings of this argument are obscured. Yet, scientific knowledge is never produced in a vacuum. The cultural beliefs of the day, prevailing scientific paradigms, and funding opportunities and constraints all play a significant role in shaping which scientific questions are answered, and which problems are pursued. In many ways, this project demonstrates the impossibility of understanding population-environment or population-climate science outside of the context of politics. Because the *perception* of demography, ecology, atmospheric sciences, and other quantitative sciences is that they are objective, rational, and value-free, the ease with which political values are embedded in science makes this arena of advocacy attractive to potential advocates. In bringing these values to the surface, my aim is not to expose evidence of "bad science." Rather, it is to interrogate the ways politics are deeply entrenched with scientific paradigms, operating largely behind the scenes.

Second, I argue that populationism is currently growing in popularity because of its appeal to young people, who are attracted to the neoliberal language of individual choices and consumer actions as solutions to large-scale environmental problems. This approach infuses how youth activists understand the concept of women's empowerment: it is action-oriented and facilitated through individual access to contraceptives, based on narratives of individual personal responsibility. These framings are key drivers of young people's desire to be population advocates as well: the narrative of the individual, action-oriented development actor infuses trainings that construct youth advocates as development experts and leaders on a global stage. The private consumer choice model is so pervasive as a part of American culture that it deeply informs our ideas about morality, individualism, and personal responsibility in a range of ways—including how we think about reproduction and environmentalism. While the young activists I interviewed proclaimed that broad social values such as women's empowerment and reproductive justice are their primary motivations for their advocacy, I demonstrate throughout the book that what actually resonates with them most is a neoliberal activist model focused on individualism.

Third, and most importantly, the term "social justice" seems to be losing its meaning. Drawing on language and strategies rooted in civil rights activism, a wide range of social actors describe their work as social justice–oriented. This has been an effective and attractive rallying cry for young people who are leading major movements on issues ranging from racial justice (Black Lives Matter) to wealth inequality (Occupy Movement). However, the vagueness and lack of precision of the term has led many other kinds of advocates, including those representing conservative right-wing social movements, to adopt this language, from the Tea Party to the pro-life movement.[10] As this terminology becomes diluted, anyone can lay claim to it—including populationists whose advocacy has uncomfortable historical resonances with population control. In so doing, they potentially undermine the work of more radical movements that truly support reproductive justice by challenging the structures of inequality that have shaped reproductive politics historically, and continue to do so today.

Population and Climate Change in Development

From 2009 to the end of 2010, I followed a loosely assembled network of international development actors whose goal was to increase U.S. funding support for international family planning policy by emphasizing its benefits for climate change, SRHR, and poverty alleviation. They pursued this goal through workshops, advocacy training sessions, research reporting sessions, private meetings, and Congressional lobbying visits. Members of the network came from a range of backgrounds. Some were donors at private philanthropic organizations or at the U.S. Agency for International Development (USAID); others worked at NGOs, universities, or were simply volunteers in local communities or on college campuses. Some identified as feminists and reproductive health advocates, while others were almost entirely consumed by environmental concerns, and saw the importance of women's reproductive health and rights as secondary. As one network member stated in an interview, "There is no one uniform movement; there are several streams or groups. Some are more concerned with demography, some family planning, some the union of environment and reproductive health concerns." Despite this diversity, what held the network together was the belief that providing poor women in the Global South with contraceptives will slow global population growth, and that this in turn slows the pace of greenhouse gas emissions and helps countries adapt to climate change, eventually benefiting us all.

There are two key challenges to this perspective, the first being that the countries that contribute the most greenhouse gases to the atmosphere every year are not those with the highest fertility rates. China, the U.S., Russia, India, and Japan emit the highest greenhouse gases (the gases that trap heat in the atmosphere) every year; each of these countries has low fertility rates (measured as the number of children born per woman over the course of her lifetime). The average woman in China gives birth to 1.5 children over the course of her lifetime. For the U.S., India, Russia, and Japan, the numbers are 1.8, 1.7, 2.4, and 1.42, respectively. Compare these numbers with the countries population advocates focus on for their family planning advocacy efforts: high fertility countries in sub-Saharan Africa like Nigeria, Tanzania, and Ethiopia. In those countries, fertility rates are 5.6, 5, and 4.4 respectively, while their national greenhouse gas emissions (GGEs) don't even rank in the top 75. In fact, with the exception of South Africa (which has a fertility rate of 2.3), none of the countries in sub-Saharan Africa rate in the top 75 greenhouse gas–emitting nations.[11]

The second challenge is that fertility rates are coming down all over the world, and they have been for the past fifty years or more, albeit unevenly. According to a 2015 report by the United Nations that looked at data from 197 nations, the average woman worldwide has 2.5 children in her lifetime, a steep drop from rates of the past.[12] Women in most parts of the world get married later, have fewer children, and access higher education at higher rates than ever before. At the same time, the report found that in 2014, 145 million women around the world had an unmet need[13] for family planning—a measure of the number of women who would like to delay or stop childbearing, who are sexually active, but who are not using Western contraception.

In these figures, sub-Saharan Africa stands as an outlier—which is why the continent receives the lion's share of focus in populationist circles. Approximately half of women in sub-Saharan Africa who would like to access Western contraceptives report not having full access. The region has the world's highest fertility rates, the lowest contraceptive coverage, and among the lowest ages at marriage. These trends are changing, though somewhat slowly. Fifty years ago, the average woman in sub-Saharan Africa had 6.7 children, compared with today's 5.1. In addition, these figures mask wide variation between countries on the continent, where fertility rates are impacted by religious and cultural diversity, poverty, social inequality, and uneven access to health care and education. As women's access to education increases along with the growing trend toward urbanization, fertility rates are expected to continue to decline across the continent.[14] Average fertility rates across the continent are projected to reach 3.9 children per woman by 2030, and 3.1 children per woman by 2050.

The significance of African fertility trends must also be understood within the context of broader global discourses on population, particularly those framed through the language of crisis. If Western contraceptive use, replacement fertility (two children per couple), monogamous, married couples, and educated, working women in urban settings are the modern ideal in development narratives, discourses on Africa stand in sharp contrast to that. The continent is all too often reduced to narratives of endless failure, poverty and disease, war and extreme inequality, described as always in need of Western intervention and salvation. These racialized, colonial narratives have long constructed Western imaginaries about the continent: Western societies have long viewed Africa as their "radical other," a counterpoint to "their own constructions of civilization, enlightenment, progress, development, modernity, and . . . history."[15] As novelist Chimamanda Ngozi Adichie reminds us, circulating a "single story" of a place, or a people, is dangerous.[16] The elisions and erasures required to create such a limited narrative not only reinforce enduring stereotypes, they also circumscribe the imaginations of those who construct and maintain the narrative, in ways that foreclose new knowledge and forms of engagement. These limitations directly shape understandings of African social, cultural, and gender relations governing sex, fertility, and reproduction, locking the continent into static images of high fertility in a fixed, unending pattern of explosive growth.

Part of the reason for this is because population-climate narratives tend to naturalize poverty and inequality in the Global South. In advocacy trainings, images of the poor are presented as dark-skinned women of color, often in tattered clothing and surrounded by children. While these images do reflect the lives of some women, they are also constructed representations, informed by colonial legacies that circumscribe Western audiences' understandings of the lives and experiences of those depicted. Dogra's 2011 analysis of images used in international non-governmental organizations' (INGOs) advocacy and fundraising materials demonstrates that depictions of racialized Global South women are used to project "universal" values of motherhood and womanhood at the same time that they symbolize Global South difference. Stylized images of women and children are particularly common in depictions of disaster, famine, the environment, and grinding poverty, presenting visual narratives of women and families—devoid of men—in ways that represent women as vulnerable and needy. Their roles as mothers are represented as universal and homogenous, in contrast with missing men, whose absence can be read as either lack of family presence and responsibility—or worse, that missing men are the cause of women's vulnerability.[17] These images of excessive fecundity, lack of stable male presence, and extreme poverty pathologize poor families while essentializing women's vulnerability. More importantly, they render women as always in need of, and in their vulnerability, always deserving of, Western assistance.

As African feminist Everjoice Win argues, the image of the African woman as "always poor, powerless, and invariably pregnant, burdened with lots of children, or carrying one load or another on her back or her head . . . is a favourite image, one which we have come to associate with development. Like the fly-infested and emaciated black child that is so often used by international news agencies, the bare-footed African woman *sells*. Without her uttering a word, this woman pulls in financial resources."[18] While this image proliferates, alternative images are obscured, particularly those of the rapidly growing urbanized African middle classes. In rendering class diversity invisible, the needs and desires of middle class African women are erased, foreclosing opportunities for development policies that address gender inequality across class lines.

Ironically, similar images of women of color in the U.S. historically accomplished just the opposite. Prior to the mid-1960s, American news media represented poverty in images of white families, primarily in Appalachia. However, in the two-year period from 1965 to 1967, these images became dramatically darker—with images of African Americans coming to dominate the "face of poverty" in the news—while poverty rates remained stable. This dramatic shift was facilitated by a change in the moral tone through which poverty was described. As coverage of the poor was presented in a more critical light and narratives of the "undeserving poor" proliferated, images of African Americans as representative of "the poor" flourished as well. Further, as poverty came to be seen through a less sympathetic lens in the news, it was increasingly associated with blackness, undeserving-ness, and laziness.[19] Over the next decade, these images of the black, undeserving poor coalesced around images of single motherhood, closely associated with Ronald Reagan's (fictional) model of the welfare queen: the woman who gamed the system, whose children were produced as a strategy to drain state resources and put cash payments into the pockets of women who did not work. The idea of blacks as inherently poor and undeserving—and of black women in particular as continually reproducing a race of resource-draining children—was central to the policing of black women's bodies in the second half of the 20th century, particularly the coercive sterilization of women of color in the U.S. (see chapter 5).

Yet, simplified images of conditions in the Global South remain central to the visual narrative of development, tugging at the heartstrings of audiences in the Global North, who project their own values onto the difference rendered in these images.[20] This visual imagery is central to producing and maintaining ideas of sexual stewardship. As Chandra Mohanty famously described, the homogenous, universal Global South woman in these and similar images "leads an essentially truncated life based on her feminine gender (read: sexually constrained) and her being 'third world' (read: ignorant, poor, uneducated, tradition-bound, domestic, family-oriented, victimised etc) . . . in contrast to the (implicit) self-representation of Western women as educated, as modern, as having control over their own bodies and sexualities, and the freedom to make their own decisions."[21] Sexual stewardship discourses are similar: Global South women are presented as the homogenous poor, simultaneously burdened by environmental problems, poverty, and their own excess fertility. These representations trap them in a static, unchanging narrative of victimhood and vulnerability that contrasts with the potential for assumed agency through reproductive self-management. This model, which depoliticizes and dehistoricizes poverty, gender inequality, and environmental problems, is the basic recipe for sexual stewardship.

This book is written as a cautionary tale against the ways contraceptives and family planning are taken up for advocacy in international development by environmental activists. I argue that advocacy strategies that problematize global population growth as an environmental and climate problem, and contraceptives as its solution, are often simplistic and do more harm than good. While family planning advocacy is a laudable goal that needs more support and increased funding, the narratives that have been used by the Sierra Club and their partners have reproduced problematic ideas about demographic change as the driver of environmental problems in ways that obscure social, political, and economic causes. They rest on an "apolitical ecology"[22] approach that says environmental problems are caused by forces such as individuals' poor choices, and by demographic changes including population growth. This approach leaves no room for understanding structural, political, and economic drivers of environmental change, and in obscuring these forces threatens to undermine the very goals populationists claim to hold dear: social and ecological justice and bodily autonomy for poor women in the Global South. Apolitical ecology approaches have long been at the heart of struggles to make sense of environmental problems—and have been central to the history of populationism and population control.

De-Naturalizing Environmental Problems

A key argument of this book is that population-environment discourses and logics persist in international development due to the enduring power of Malthusianism. Yet, Malthusian logic has been thoroughly critiqued—many would say debunked—by critical scholars and activists for decades now. Many of the most trenchant critiques of Malthusianism have been offered by scholars of political ecology. For the past forty or so years, combining "the concerns of ecology and a broadly defined political economy,"[23] political ecologists have rejected simplistic biological, demographic, and other apolitical framings of problems like deforestation, soil erosion, and climate change, instead rooting these problems in unequal social, political, and economic systems.

Beginning in the 1970s, Marxist political ecology writers have analyzed the role of class and capitalism in Malthusian logic. Harvey, for example, argued that the concepts of nature and natural resources are not fixed, but rather are constructed socially through capitalist systems that assign them value. Resource scarcity, then, did not arise from biological conditions of overpopulation or human overconsumption, but instead from outcomes of inequitable distribution of wealth.[24] Studies of local and regional famine events demonstrate that international markets and capitalist systems of production, not overpopulation, have produced entrenched hunger and poverty at the local level.[25] For example, Davis's analysis of nineteenth-century regional famine events found that these famines, long constructed in neo-Malthusian scholarship and media accounts as the result of local overpopulation and unsustainable cultivation practices, were actually the result of both of El Niño weather patterns that shape long-term trends in soil fertility, as well as the introduction of colonialism and capitalism in Global South countries. The resulting entrenched poverty and maldistribution of food resources, based on the deep political-economic inequalities of colonization, manifested in widespread starvation at national and international scales.[26]

Political ecologists have also critically interrogated knowledge production itself, specifically the ways scientists make and circulate paradigmatic ideas about environmental change. Fairhead and Leach's sharp analysis of degradation narratives describing Guinean forest and savannah landscapes details the ways ecological scientists and policymakers literally saw population-driven deforestation where it did not exist, by misreading islands of dense forest cover in broader savannah wildlands as evidence of *deforestation*, when in fact they exemplified *afforested* landscapes, restored by local communities. The scientists, presented with data counter to the neo-Malthusian narrative so deeply embedded throughout the ecological sciences, had no other framework through which to interpret and explain what they were seeing, and thus produced decades of erroneous results.[27] In a similar vein, Jarosz's historical study of deforestation in Madagascar demonstrates that, counter to prevailing arguments, the dramatic loss of forest cover across the island was driven by colonial policies favoring market-based production of forest commodities, rather than population growth. In fact, her research found that, under colonial rule, forest loss was intensified during periods of population *decline* in the country's eastern rainforest corridor.[28] Numerous studies have demonstrated that the intersection of population control programs and rural development schemes have resulted in the *intensification* of local poverty, land dispossession, and few, if any, impacts on fertility trends.[29]

Ecological problems, and the solutions that are devised to respond to them, do not simply exist in the world of their own accord. Rather, they are constructed by assemblages of actors and knowledge practices. Hajer illuminates this through his analysis of discourse coalitions. Discourses, or "ensemble[s] of ideas, concepts, and categories through which meaning is given to phenomena,"[30] provide the tools for constructing problems, as well as providing the context in which problems are understood. A discourse coalition—comprised of actors who share a given social construct—then takes these up, using persuasion or force to convince others to accept their interpretation of reality. When specific problems are discussed, discourses are presented as particular narratives or storylines; the complexities of the various narratives are concealed as they are assembled into a coherent whole. Policymaking is similarly discursive; it is not just about problem solving, but rather also about problem creation. This does not deny material reality, but rather argues that environmental problems, for example, cannot be understood without analyzing the discursive practices through which we perceive reality and the options available for intervening on it.[31] Further, discourse coalitions function as advocacy networks.[32] These networks provide non-traditional actors the ability to mobilize information strategically and transform the terms of policy debate in ways that influence more powerful institutions.

Thinking through these issues with respect to population and reproduction necessitates an analysis of questions of gender, and its attendant roles, norms, expectations, and inequalities, all of which impact environmental discourses, policies, rights, and access to resources. Feminist political ecologists have long argued that experiences, interests, and responsibilities for nature and environment are often constructed along the lines of gender inequality, and mediated by race, culture, and gender.[33] Their work demonstrates that gender operates as a critical variable in shaping resource access and control, interacting with other social categories such as class, race, culture, and ethnicity to shape processes of ecological change, struggles to sustain ecologically viable livelihoods, and sustainable development. However, feminist political ecology does not simply add another category (gender) to political-economic analyses of environmental change; rather, it also addresses questions of identity and difference, and how multiple forms of meaning are made in relation to environmental struggle and change.

Moreover, feminist political ecology rejects the essentialist narratives, common in gender, environment, and development scholarship of the

1980s, that posit women's relationships to nature and the environment as homogenous, static, and universal. Gendered social *roles* shape environmental practices, policies, and their impacts, manifesting as unequal norms, burdens, expectations, and blame narratives. Women's employment in formal and informal sectors, roles within environmental justice organizing and other social movements, and centrality to smallholder farming, traditional medicine gathering, and water and fuel collection have all positioned women within close proximity to nature. This is not based on a natural affinity for, or inherent closeness to, nature, but rather the demands, expectations, constraints, and opportunities shaped by social norms and expectations. While much of the literature on gender and environment is focused on the Global South, feminists have also analyzed these relations in the Global North, particularly with respect to how gendered social norms impact household resource consumption and use. For example, when researchers and policymakers focus on "green duty" lifestyle changes—encouraging consumers to reduce resource consumption, to use green products, and to educate themselves about the impacts of everyday household products—these efforts fall heavily on women's shoulders.[34]

An enduring storyline of gender and environmental change has predominated in environmental policymaking for decades.[35] In this storyline, women are vulnerable victims, subject to the harsh impacts of environmental changes, based on an assumed close relationship to nature. At the same time, development projects characterize women as particularly resourceful and able to adapt to environmental changes. Thus, a second image arises: that of the resilient and responsible actor, well poised to take matters into their own hands to turn environmental problems into environmental solutions.

This narrative, of closer-to-nature, victim, and potential agent of change, has characterized much of the environmental thinking on women in the environment and development sector. Resurreccion[36] posits these enduring storylines, or "persistent women-environment linkages," as an important historical means for women to secure a seat at the table in terms of environmental policymaking. In order to play a role or have their concerns recognized at all, women historically had to simplify and adopt essentialized representations. However, these narratives are also detrimental: they cast women in narrow, homogenous roles that offer little room for diversity, historical contingency, or innovation. They also erase men, and therefore gender itself as a relational condition of social inequality, from the picture. In so doing, these representations depict women's condition as static and unchanging, thus naturalizing vulnerability rather than rooting it in social relations.

Ironically, the images of victims-and-agents arose from efforts to better understand women's particular roles in environmental management in the Global South. In the early 1970s, development debates arose around the linkages between Women, Environment, and Development (WED), and focused on addressing "all women's interrelations with the environment in the context of economic development as well as the effects that environmental degradation has had upon women's lives,"[37] including increased subsistence household labor (water and fuel collection), effects of pollution of air, water, and soil, and increased workplace exposure to toxins. Braidotti, et al. locate the emergence of WED debates in forestry (specifically fuelwood energy) and agriculture circles in early questions around fuel and water use, and women's role as collectors of natural resources for household consumption. Women began to be framed as unique victims of energy and environmental crisis, and the category "women" in this context began to be used interchangeably with "the poor." At the same time, international development circles began to focus on women's grassroots environmental activism in the Global South, concentrating on the Chipko Movement in India and the Greenbelt Movement in Kenya as examples of women's particular role as protectors of the environment.[38] However, the broader engagements with women's roles in, and relationships to, development extend far beyond questions of the environment and natural resources—they range back to the earliest days of U.S. involvement in international development intervention, and shape how women, gender, and gender relations have been conceptualized and made into development intervention strategies.

Untangling Women, Gender, and Empowerment in Development

Women in Development (WID) strategies were created in the 1970s in an effort to integrate women into existing international development initiatives focused in the political, economic, and social sectors. Following Ester Boserup's groundbreaking work on the sexual division of labor in agrarian communities, women development practitioners in Washington, DC, were concerned with ensuring that women would be better integrated into local economic structures in the Global South.[39] Their perspectives were informed by, and closely associated with, modernization theory—the paradigm that dominated international development from the 1950s to the 1970s and focused on technological fixes such as technology transfer, market-based skills, and the development of technologies to decrease or better marketize women's workload.

Feminist critics of WID approaches charged that they were based on an acceptance of existing, unequal social, political, and economic structures, and avoided questioning the reasons for women's subordination. They also focused on "women" as a singular and homogenous group, thus obscuring the role of unequal relations with men, as well as the importance of race, class, ethnicity, and religion in shaping women's life conditions. Given that WID programs were primarily focused on how to integrate women into existing development initiatives, another central assumption was that women were not already participating in development programs, which was far from the case. In fact, women were first included in international development in the 1950s and 1960s through a focus on their reproductive roles in the household, via programs and policies addressing food aid, malnutrition, and family planning. The approach primarily focused on women's role as mothers, with an underlying rationale of social welfare.[40] International economic aid prioritized male-dominated industries in the formal sector, while welfare for the family targeted women. This welfare approach was based on three assumptions: first, that women are passive recipients of development, second that motherhood is the most important role in women's lives, and third, that raising children is the most effective role for women economically. In other words, women's primary roles and contributions were assumed to be reproductive, while men's roles were productive. As Moser argues, "Intrinsically, welfare programmes identify 'women' rather than lack of resources, as the problem, and place the solution to family welfare in their hands, without questioning their 'natural' role."[41] Over time, development program managers argued that the competitive free market was a better venue for maximizing women's potential via opportunities for self-improvement. As a result, women's roles as economic agents (micro-entrepreneurs, farmers, factory workers) were increasingly recognized as central to integrating women into development.[42]

In the early 1980s, feminists mounted a formal response to the embedded power relations and inaccurate assumptions of WID programs, under the banner of Gender and Development (GAD). GAD approaches address not only women, but women in relation to men, and how gender relations are socially constructed within the contingent contexts of space and time. In the GAD approach, both production and reproduction are socially constructed, and both create the conditions of women's subordination—thus social, economic, and political life are all sites for questioning and critiquing the roots of women's subordination in the context of socially constructed gender roles. GAD proponents were particularly concerned with integrating these critiques into development approaches, arguing that women were active participants in development from the beginning, operating as agents of change rather than passive recipients of outside interventions.

A central concern within GAD is the question of power—how it operates, circulates, and is embedded in development. Empowerment is central to these concerns, particularly transforming power relations within development. GAD proponents have developed multiple ways of thinking about empowerment as a resource to transform interventionist models from a grassroots, bottom-up perspective, as well as altering understandings of what power is, how it operates, and how it can be harnessed and transformed.

For example, Kabeer identifies three ways of thinking through empowerment: the "power to" affects outcomes over and above the wishes of others; "power over" refers to procedures that benefit certain groups at the expense of others; and "power with" focuses on building solidarity and alliances with others.[43] Sen and Batliwala extend these orientations in multiple directions, defining empowerment as the ability to mobilize resources and to determine the rules of the game in ways that mask the workings of inequality; as a "process of changing power relations in favour of those at the lower levels of a hierarchy,"[44] and as control over both material resources and ideology. Sen and Grown, however, insist that women's empowerment must be an explicitly feminist enterprise that responds to the needs and priorities of multiple kinds of women, and that is "*defined by them for themselves*" (emphasis in the original).[45] Their vision of empowerment within development prioritizes strategies to meet human needs, as well as transforming access to and control over economic and political power. In this model, improved living conditions, socially responsible resource use, and the elimination of gender subordination and socioeconomic inequality are all linked sites of struggle and necessary transformation.[46]

However, as Halfon's analysis demonstrates, development institutions are generally not organized around grassroots work, and tend to function in more top-down structures. They also emphasize program efficiency based on short term, easily measurable goals. Feminist empowerment models are rooted in radical grassroots political change, which does not fit easily with the kinds of bureaucratization and hierarchy found in many development institutions. In the context of population, many NGOs narrow their focus to reproductive decision-making as the center of empowerment, likely because this approach serves as a discursive link between traditional populationist groups and women's rights advocates.[47] As empowerment became incorporated into population policy in the 1990s, it was translated from a political approach to social transformation into a set of policy strategies. This translation effort has diluted its radical framing in favor of efficiency and achievement of development program goals—ironically, this is the very kind of approach the empowerment model was designed to critique and resist. Halfon describes the process of dilution thus: "The meaning of women's empowerment has become defined through the practices and discourses of population institutions rather than strictly through feminist and radical theorizing. Because of the looseness of its definition, its reinterpretation through existing policy goals and planning strategies, and the constraints posed by institutional and professional needs, empowerment-oriented projects often resemble the development frameworks they were originally conceived in opposition to."[48] As I argue throughout this book, empowerment has become such a loose term that it is rather far afield of its original intent. Today, in population circles and in development discourse more broadly, empowerment is often "used in a way that robs it of any political meaning, sometimes as no more than a substitute word for integration or participation in processes whose main parameters have already been set out elsewhere."[49]

Sex, Population, and Science in Development

As Foucault has argued, population became an economic and political problem requiring state level management and surveillance in the eighteenth century. This was not a problem to be solved through elimination, but rather one to be managed and governed in order to optimize the capacities of populations. Techniques such as statistical data collection, public health surveillance, collection of information on birth and death rates, fertility, life expectancy, and health and illness status came to operate as an opportunity for state authorities to direct the sexual energy of populations, serving as a point of entry for ever-increasing interventions into the most intimate spheres of family life. Sexual and reproductive health interventions, including family planning, population control, and policies designed to increase birth rates have all come to be managed in this way over time. Sex and sexuality, reproduction, and life itself are objects of study and intervention precisely because of the power relations governing knowledge production, surveillance, and management of populations and bodies that made them possible as such objects. And the knowledge and discourses of sex and reproduction are deeply invested in questions of power, authority, and sovereignty.[50]

Discourse operates as a vehicle for the consolidation of power based on the management of bodies, sex, and reproduction. Ideas of nation, state, and progress have long been tied directly to sexual conduct and reproduction through science, modernization, and capitalism—and development interventions, particularly in the arenas of sex and reproduction, operate as vehicles for such projects. As sexuality has historically been transformed into an object of scientific and medical study and intervention, sexual and reproductive health development programs have advanced notions of what it means to be modern in one's sexual and reproductive behavior. Such programs make possible new forms of governance in which bodies and health become sites for control, management, surveillance, and dominance. In this context, sex operates as a moral object as well as a site around which projects of standardization, universalization, and scientific transformation are organized. In the context of development, sex is simultaneously rendered a moral object, an object of scientific knowledge production, and a site for expanding notions of progress and modernity.[51]

Population advocates in the development arena have long been in a constant quest for more scientific data to prove the link between population growth and environmental change, to demonstrate that their advocacy is driven by factual, unbiased science. The idiom of co-production, conceptualized by Jasanoff, is particularly illuminating here. Jasanoff argues that "the realities of human experience emerge as the joint achievements of scientific, technical and social enterprise: science and society, in a word, are *co-produced*, each underwriting the other's existence."[52] Co-production arises from the recognition that "the production of order in nature and society has to be discussed in an idiom that does not, even accidentally and without intent, give primacy to either," and as a result is reflective of a "self-conscious desire to avoid both social and technoscientific determinism in S&TS (science and technology studies) accounts of the world."[53] It does not conceptualize truth and power as pre-formed entities that oppose each other, but rather argues that scientific knowledge and political orders "shape, entail, and refer to each other."[54] These are also powerful processes: the ability to produce, shape, and circulate knowledge is deeply linked to notions of authority and expertise, the hidden practices of which co-productionist analyses help to expose.

While population-environment advocacy has been grounded by enduring Malthusian ideologies, shifting political climates have forced the development of new discourses, frameworks, and ideological approaches. Analyzing the relationship between science and politics helps account for these shifts, even as population sciences continue to proliferate. One of the key aims of this project is to explore the political context of scientific knowledge production around population and the environment. This is not to rehearse debates about the role of "objectivity" in science, but rather to demonstrate the social and political contexts within which scientific knowledge is produced, and within which it continues to be deeply entangled. In investigating these questions, I aim for this project to serve as a corrective to ideas that naturalize relationships between population growth and environmental problems as fixed, linear, and apolitical reflections of the material world.

Methodological Entanglements

My research initially began with an exploration of the institutional politics of environmental non-governmental organizations (ENGOs) working at the intersections of global population, environment, and sexual and reproductive health (SRH) in international development. The ENGOs that have been active in this arena include the Sierra Club, as well as the National Audubon Society, Worldwatch Institute, World Wide Fund for Nature, Conservation International, and Population Connection (formerly known as Zero Population Growth; despite the fact that the organization's name is population-focused, a senior manager described it in an interview as an environmental organization). Most of these organizations employ a small staff, often designating one employee to serve as their "population person," responsible for informing NGO members about global population trends, tracking U.S. legislation on international family planning, participating in congressional lobbying and other policy advocacy, presenting their work at conferences, and when possible, joining with other members of the population advocacy network for activities.

Over time, I realized that population advocacy at these institutions is constituted by relations within a network of similarly engaged actors, from individual donors to community and campus activists. As a result, my ethnographic lens was retrained from a focus on ENGOs to a focus on the network itself, studying how youth activists come to be so deeply imbricated within it that they see its aims and priorities as their own. A key aim of this book is to track the practices through which youth and others in the network deploy political strategy and scientific knowledge on global population growth and environmental change to produce, circulate, and ground new modes of policy advocacy. These practices are far from static in their development—rather, they are the result of the careful and persistent efforts of network members who view population interventions as necessary to ensure environmental sustainability at local and global scales.

Throughout this book, I refer to this group alternately as population-environment advocates or simply population advocates. Through interviews and participant observation with youth and other advocates, NGO program managers, and donors, I sought to understand the politics and practices of international development policy advocacy from the perspective of those working behind the scenes. One of my goals here is to demonstrate the multiplicity of motivations, goals, ethical positions, and moral frameworks utilized by those in this field of development, as they attempt to advance a coherent movement while navigating the thorny terrain of controversy. A secondary goal is to investigate the ways in which knowledge about the body, particularly poor women's bodies and fertility in the Global South, is constructed, disseminated, and utilized as the basis for political action by a network of actors located at vast geographic, cultural, political, and economic distances from those whose experiences they claim to represent. In other words, I attempt to understand how the "other," in this case the universal "Woman" of the Global South (Mohanty 1992), is constructed through the melding of scientific data and social activism in order to advance an international policy agenda.

My research is grounded in three modes of data collection: participant observation at population advocacy trainings, workshops, research presentations, and conference sessions; in-depth interviews with members of the network and other activists; and analysis of archival documents. I conducted fieldwork that took place over a twenty-one-month period from April 2009 through December 2010. During this period, I attended a dozen workshops, conferences, trainings, and research presentations that addressed population growth and environmental change. Of these meetings, most advocated for reducing global population growth in order to promote environmental and climate sustainability. One conference took a distinctly different approach, using a critical race and gender analysis to reject neo-Malthusian arguments in favor of an approach centered on reproductive justice. Many of the population-environment advocacy trainings I attended were focused primarily on enrolling youth activists from campus-based environmental and SRH clubs at colleges and universities around the U.S. These multi-day trainings were primarily led by the Sierra Club, in conjunction with a series of SRH and women's advocacy organizations, ranging from the Feminist Majority Foundation to the International Women's Health Coalition, and took place in Los Angeles, Washington, DC, and San Francisco. Attending trainings also provided the opportunity to conduct in-depth interviews with youth population advocates, as well as ENGO and SRH NGO representatives.

Near the end of my fieldwork, I spent three weeks in Cancun, Mexico, participating in the international climate change conference (also known as the 16th annual Conference of Parties, or COP 16), and the associated youth-led Conference of Youth (COY6). Both provided key opportunities to observe how youth activists trained by the population-environment advocacy network operationalize their advocacy training in an international context. Over the course of the project, I conducted formal, one-on-one interviews with sixty-four NGO representatives, donors, community activists, scientists, and scholars. Approximately twenty additional interviews were conducted with feminist activists who critique the network's strategies from an intersectional perspective focused on analyzing the race, gender, and class politics of the network's efforts. Finally, I pored over dozens of program reports, project descriptions, funding analysis documents, meeting notes, and funder network reports, in order to supplement my ethnographic material with archival and contemporary documents. My informants were very generous in providing me with these documents, as well as sharing private correspondence.

On Current Politics

The stakes of writing a book like this one are high, particularly in the current political moment. As I write this, Donald Trump, a reality television star who has been accused of multiple sexual assaults, who has repeatedly questioned the existence of global warming, and who has referred to violent white supremacists as "good people," is President of the United States. In his early days in office, he held the world in limbo for months, refusing to come to a decision about whether the U.S. would remain party to the Paris Agreement—the first legally binding climate change agreement in which all signatories have agreed to work together to limit global temperature increase to less than two degrees Celsius (he eventually made the decision to withdraw the U.S. from the agreement). Just two days after the largest women's march in history, he reinstated the Global Gag Rule, an order that prevents NGOs outside the U.S. from receiving U.S. family planning funding if they provide abortions, or even offer counseling, referrals, or educate clients on obtaining abortions elsewhere. In a particularly shocking and unprecedented move, at the time of this writing, the Trump administration had just delivered a proposed budget for fiscal year 2018 that eliminated all global health funding for international family planning and reproductive health. From over $600 million dollars in 2017 to zero, with a stroke of a pen.[55]

Given the funding landscape and the realities of Washington politics, population advocacy has become increasingly important from a pragmatic perspective, in order to help ensure that women around the world will have access to contraceptives and other reproductive health services. However, a critical, feminist corrective to the narratives that sustain this advocacy work is increasingly needed. This book offers this kind of corrective: it interrogates the development and enduring roles of Malthusianism and populationism in the ways activists think about, and act on, population and family planning through international development. It centers critiques of race, gender, and class politics in the construction and circulation of populationist framings, and the ways activists link environmental degradation and climate change to human numbers. It also explores the politics of knowledge production and the stakes of the close relationship between science and policy advocacy in creating and sustaining international family planning advocacy efforts. With this said, I have long been, and remain, a strong advocate for women's voluntary access to contraceptives and other family planning services within a context of comprehensive sexual and reproductive, and broader, health services.

Organization of the Book

This chapter has laid the groundwork for understanding the conceptual questions that this book seeks to address. In chapter 1, I take up the tumultuous politics of population in development, linking historical controversy and international policy activism with the context of declining donor funding over time, and the recent focus on population as a climate problem. I argue that, as climate change discourses bring environmental crisis narratives to the forefront of population debates, these narratives open the possibility for new ways of considering reproductive governance and surveillance—and a return to population control. Chapter 2 traces the history of neo-Malthusian thought, exploring how population was constructed as a scientific and environmental problem in the U.S.

over the course of the twentieth century as it entwined with American anxieties over nation, geopolitical stability, and the global racial balance of power. Chapter 3 explores the science-policy interface through an analysis of the close relationships between climate-population scientists and their funders. It analyzes the workings of donor advocacy behind the scenes, arguably the most important element in how scientists' models not only *project*, but also *produce*, potential futures.

Chapter 4 departs from a focus on knowledge production to interrogate the process through which young activists transform themselves from local policy advocates to self-styled development leaders, experts, and social justice actors. Focusing on training workshops, I explore how young people negotiate their own ambivalent subject positions as population advocates, attempting to reframe their role within the language of social justice. Chapter 5 investigates the ways mainstream reproductive health NGOs draw on the language of reproductive justice to frame population advocacy as socially progressive, while obscuring the intersectional politics that structure the reproductive justice movement's history and current work. The final chapter concludes the book not by offering solutions, but remaining in a thorny, uncomfortable, and vitally important space—a space of fretting over the future of populationism, and its ever closer engagements with gendered neoliberal development discourses.

1

The Population "Crisis" Returns

In 2010, *Mother Jones* magazine ran a special issue with a cover asking, "Who's to Blame for the Population Crisis?" Inside, the issue's lead article, "The Last Taboo: What Unites the Vatican, Lefties, Conservatives, Environmentalists, and Scientists in a Conspiracy of Silence?" centered on a crowd thronging the streets of Calcutta late one night:[1]

> It's midnight on the streets of Calcutta. Old women cook over open fires on the sidewalks. Men wait in line at municipal hand pumps to lather skin, hair, and lungis (skirts), bathing without undressing. Girls sit in the open beds of bicycle-powered trucks, braiding their hair . . . grandfathers under umbrellas squat on their heels, arguing over card games, while mothers hold bare bottomed toddlers over open latrines . . . Many people sleep through the lively darkness, draped over sacks of rice or on work carts full of paper or rags or hay. Groups of men and women, far from their home villages, sprawl haphazardly across the sidewalks, snoring.[2]

Rather than a conspiracy of silence, a review of media and popular writings at the turn of the millennium reveals just the opposite. In fact, the *Mother Jones* issue appeared in the midst of a veritable explosion of articles, blog posts, journal publications, and think pieces linking women's fertility, global population growth, and the environment. From 2005 to 2008, writing that cited the terms "population growth" and "environment" or "climate change" increased fourfold.[3] Like *Mother Jones*, magazines began to publish special issues and entire series dedicated to the issue. *Scientific American*'s population issue in summer 2009 claimed that "Malthusian limits are back—and squeezing us painfully,"[4] while the next year, *National Geographic* initiated a yearlong series focused on population growth's environmental consequences, titled "State of the Earth 2010." With striking, glossy images of people around the world farming, fishing, driving cars, extracting minerals, and thronging the streets, juxtaposed with pictures of the denuded, heavily plowed or constructed earth, the images told a gripping story of a planet sagging under the weight of human activity. And in 2011, after the United Nations published new global population projections, the *New York Times* responded by publishing articles and blogs detailing the ways ongoing population growth would intensify demands on the global food supply, colliding with limiting factors like climate change, water scarcity, and land shortages.[5]

Increasingly, these writings focused on climate change.[6] A 2015 editorial in the *Los Angeles Times* argued that "sensitive subject or not, the reality is that unsustainable human population growth is a potential disaster for efforts to cut greenhouse gas emissions,"[7] while the *Huffington Post* reported that "access to voluntary birth control . . . will cut our collective human carbon footprint."[8] Specialized scientific reports have also taken up this theme, with the Intergovernmental Panel on Climate Change (IPCC), a transnational body of climate scientists and policy actors, stating in its 2014 assessment report that "globally, economic and population growth continued to be the most important drivers of increases in CO_2 emissions from fossil fuel combustion."[9]

If population growth is a secret, it is an open one—yet, the language of secrecy and taboo continues to predominate. Collectively, the majority of recent population-environment writing is focused on two central themes: population interventions as secret, controversial, or overlooked, but necessary solutions to environmental problems; and the mutual social and environmental benefits of supporting women's universal, voluntary access to Western contraceptives. These writings insist that, while taboo, population *must* be addressed if we are collectively going to make progress on climate change, and—equally importantly—that increasing women's access to contraception as a climate mitigation strategy is empowering to women. Moreover, this argument has also been taken up by prominent Western feminists, including Hillary Clinton, Gloria Steinem, and Mary Robinson, all of whom have advocated family planning and population stabilization as a necessary, women-centered climate change strategy.[10]

Behind the scenes, a different story emerges. In international development, the arena that has turned the rhetoric of population problems into intervention programs and policies, population and family planning have lost the prominence they once enjoyed. Development actors speak often of the disappearance of population, in comparison to the days when its role in international development and foreign policy seemed unshakable. Where was population before, and where did it go? Why is it "coming back" now?

This chapter explores the historical and contemporary reasons for the supposed silence and re-emergence of population in public. It begins with an exploration of the history of global population control and the resulting feminist activism that resulted in a dramatic transformation of international population policy. It then explores the often-invisible role of funding by documenting the role of private and public sector donors in mobilizing resources and securing new strategies to restore it. Finally, I close with a discussion of why climate change is such an expedient way to bring population back to public debate—and why this approach is particularly dangerous.

The American Roots of Population Control

In 1954, Hugh Moore, founder of Dixie Cup and a population activist, wrote a pamphlet called "The Population Bomb" and began distributing it through mass mailings, eventually reaching over 1.5 million people. One of those people was General William Draper, who led a commissioned study on U.S. military aid for then-President Dwight D. Eisenhower. Draper invited population experts to brief his staff, and at the end of the commission's work, they wrote a report calling for the U.S. government to fund interventions in maternal and child welfare overseas (which, although not explicitly stated, included contraception). This marked a turn in the U.S. foreign policy establishment—one in which rapid population growth was now officially marked as a security issue, and birth control was seen as a part of national defense. Draper and Moore formed the Population Crisis Committee (later renamed Population Action International), using it as a platform to advocate the necessity of population control in international aid.

Draper, Moore, and the committee were not motivated by humanitarian or public health concerns, but rather Cold War concerns about the spread of communism. Moore's pamphlet was based on Washington's anxieties over the spread of communism in rapidly growing nations of the Global South, fears that were closely aligned with concerns about continued access to raw materials, labor, and markets in nations gaining independence from colonialism. The "bomb" he referred to symbolized the potential for rapid population growth in the Third World to have the same kind of explosive and devastating effects on earth as a nuclear bomb.[11] Demographers were critical of this approach; they argued that it was dangerous to go with the "offensive and potentially controversial" proposition that population control in the Global South should be used as a tool to counter the spread of communism. However, Moore was unmoved; he wanted to influence business and government elites as well as the general public.

Population control efforts were under way across the newly independent nations in the Global South, led both by state governments and international donors, with the U.S. government playing a prominent role. International agencies began to tie food aid and government loan packages to contraceptive distribution schemes. In the mid-1960s, the U.S. government under Lyndon B. Johnson began to produce and circulate experimental contraceptives intended for mass distribution across India, including over one million intra-uterine devices. Promotional incentives such as cash payments and radios were used as part of carrot and stick incentives for community-based contraceptive providers, with support from the Indian government. After Indira Gandhi took the reins of power in India's government in 1966, she immediately intensified population control efforts across the country. Doctors were given bonus payments in exchange for reaching targets for IUD insertions and surgical sterilizations. Non-clinical practitioners were paid per service, with vasectomies paying out double the price for that of IUD insertions. Between 1966 and 1967, 1.8 million Indians were implanted with IUDs or sterilized, a number that would increase dramatically a decade later under India's national Emergency period. During the Emergency, family planning became integrated into all government offices' activities, cash payments to IUD and sterilization "acceptors" increased more than ten-fold, and compulsory sterilization was introduced as a component of state policy, often enacted in "sterilization camps." Within one year, more than eight million sterilizations, including 6.2 million vasectomies and 2.05 million tubal ligations were conducted, primarily among the poor.[12]

As historian Matthew Connelly's comprehensive study of global population control demonstrates, while India is a particularly egregious case, it is not entirely unique in the history of global population control. Food and Agriculture Organization (FAO) aid distribution in Haiti became tied to contraceptive incentive programs, and the governments of India, Singapore, and Indonesia denied housing, tax, and other benefits to parents who had more than two children. By the early 1980s, Bangladesh was the largest recipient of international population assistance, and as a result, the state imposed harsh punitive measures on those who refused family planning efforts. In one particularly horrifying example, members of the Bangladesh army rounded up hundreds of people for forcible sterilization, and food aid from the World Food Program was denied to Bangladeshi flood victims who refused to be sterilized.[13]

This is not to suggest that women in the Global South were not interested in birth control. However, even when they did attempt to secure access to contraceptives voluntarily, the lack of consistent service delivery, comprehensive options, and education about methods and their side effects often formed a formidable barrier. The emphasis on family planning in aid programs also helped deepen poverty in already-struggling nations—particularly across sub-Saharan Africa, where U.S. foreign aid in the 1980s was predicated on countries adopting structural adjustment programs tied to population control. Connelly refers to the twentieth-century escalation of global population control as a "system without a brain," characterized by policies and other interventions that gained a momentum of their own. Population control seemed to be a machine that ran itself, and it was running amok.[14]

At the same time, population has long been a site of international struggle in the development policy arena. While population control had been central to the American international development agenda from the late 1950s, leaders in Global South countries at times resisted this approach. In 1974, when the first World Population Conference was convened in Bucharest, many leaders rejected population control imposed by the North, arguing that it was a distraction from the inequalities underlying the international economic order. Instead, they pointed out the long shift in Europe from high population growth rates to very low growth, arguing that economic and social progress were responsible, not population control. Accordingly, they developed the slogan "development is the best contraceptive."[15]

Within ten years, however, many leaders' positions had shifted. At the 1984 World Population Conference in Mexico City, leaders from the Global South argued that their resources could not support unrestricted population growth, whereas the U.S. under Reagan adopted a neutral position on population growth, arguing that slow economic development in the Global South was attributable to too much state intervention, not demographic changes. The best contraceptive, according to Reagan's administration, was the free market.

Why such a reversal? The answer had less to do with actual questions of population and development, and more to do with the rising influence of religious fundamentalism in the U.S. In the early 1980s, a well-organized group of pro-life advocates who were key to Reagan's base of support had begun pressing for three things: defunding and dismantling the population office at the U.S. Agency for International Development (USAID); using population funds to support "natural family planning"; and sending a pro-life delegation to the 1984 World Population Conference in Mexico City. In response to this pressure, the U.S. sent a delegation with a neutral position on population growth; at the conference, they argued that free markets provide a solution to social problems like poverty and inequality, and that family planning alone could not address these issues effectively. Abortion could not address them at all. As a result, the Reagan administration withdrew support from any organization providing abortion services, information, or referrals (this became known as the Mexico City Policy, also known as the Global Gag Rule). In this context, the Mexico City Conference "marked the moment when population growth was no longer treated as a global problem requiring a global solution."[16]

By the time of the next world population conference at Cairo in 1994, the terms of the global population debate had shifted once again, this time to focus on the role of population in producing environmental problems. And, as the next section will demonstrate, the policy document produced at the Cairo conference reflected a perspective that had never before been formally enshrined in international population policy: a focus on women's rights to healthy sexuality, reproduction, and autonomous bodily decision-making.

The Roots of Sexual Stewardship: Rio, Cairo, and Women's Embodied Environmental Responsibility

Feminist organizing for the 1994 International Conference on Population and Development (also known as the Cairo Conference) began years before the conference. In preparation for the 1992 UN Conference on Environment and Development (UNCED) in Rio, also known as the Earth Summit, activist gathered to outline their position and organizing strategy. The Women's Environment and Development Organization (WEDO) convened a 1991 meeting of over one thousand women activists from ninety countries who gathered in Miami to discuss the "women's dimension" of the global environmental crisis. Unlike neo-Malthusians, the feminists at the conference de-centered population from the debate, instead focusing on the global economic crisis, third world debt, and the impacts of nuclear testing and other military operations on land, wildlife, and human health across the globe. A central theme of the conference was that these effects were gendered, in that they disproportionately impacted women's bodies and livelihoods. They also focused on the political dimensions, specifically the fact that women were drastically underrepresented in decision-making bodies that made global environmental policy.[17]

The following year, at the UNCED meetings, debates about the role of population growth emerged in an area of the conference's non-official NGO forum known as the Women's Tent (Planeta Femea). The Tent was co-organized and hosted by the Brazilian Women's Coalition and WEDO, and offered a program of daily workshops and presentations structured around drafting a Population Treaty and a separate Women's Treaty. By the time negotiations had concluded, coalition actors led by Southern anti-Malthusians and Northern feminists firmly rejected inclusion of population control as a component of international environmental development, opting instead for a critical focus on Northern consumption practices and advocating for development approaches favoring women's comprehensive sexual and reproductive health care.

The document produced from these debates is known as the Treaty on Population, Environment and Development. It asserted that "women's empowerment to control their own lives is the foundation for all action linking population, environment and development," and explicitly rejected all forms of control over women's bodies by governments and international institutions, including coerced sterilization, experimental contraceptive development, and denial of access to abortion.[18] Although the document located the drivers of global environment in militarism, debt, structural adjustment, inequitable trade policies, and patterns of consumption and production in the industrialized North, it did include mention of population. However, this mention of population was in sharp contrast to prevailing arguments, focusing on resource consumption in the Global North. In the preamble, the treaty's authors stated that they "affirm and support women's health and reproductive rights and their freedom to control their own bodies," an approach which demanded "the empowerment of women, half of the world's population, to exercise free choice and the right to control their fertility and to plan their families."[19] The document made it clear that population was not central to discussions of Global South environmental development and poverty reduction, aside from a critical analysis of the role of Northern populations' consumption patterns in producing degradation.

Some of the language from the treaty became the basis for negotiations among diverse actors—neo-Malthusians, feminist activists, women's health advocates, population controllers, environmentalists, and others—in preparation for the 1994 International Conference on Population and Development (ICPD) at Cairo.

In the leadup to Cairo, feminist activists from the Global South were concerned with ensuring that language on women's empowerment, sexual health, and reproductive self-determination would be included in the upcoming negotiations. Unlike the treaty produced in the Women's Tent, the document coming out of the ICPD meetings at Cairo would be enshrined as formal international policy governing how population interventions would be devised moving forward. The stakes were very high. Reproductive health activists situated contraceptives as one component in a broader constellation of reproductive health services for women, including maternal and child health care, cancer prevention, abortion access and post abortion care, and STD prevention and treatment. "Women's rights principles and respect for bodily integrity and security of the person" were also central to the framework, which offered a "groundbreaking consensus among feminists and the international community, going well beyond the basic-needs approach historically advocated by policy makers in the South."[20]

Reproductive rights became the central framework for their advocacy efforts, situated within a broader context of needs focused on women's health across the life span, and not just the reproductive years.[21] By the time the ICPD meetings began, women in the South were a leading and majority presence in transnational feminist organizing for the conference, and they had articulated a framework that situated sexual and reproductive health in a context of human rights and global economic policies. Health was integrated as a component of a broader development agenda focused on international debt, structural adjustment, and state investments in basic needs.

A Southern coalition, Development Alternatives with Women for a New Era (DAWN), played a prominent role advancing Southern feminist interests. They advocated negotiating and applying pressure to the development establishment—including the population establishment—by situating women's reproductive health within a comprehensive human development framework. In this framework, meeting the basic needs of the poor, particularly women, was necessary to empower them to control their own lives and livelihoods, and crucial to ensuring sustainable development, equitable economic growth, and human rights. DAWN also critiqued Western cultural notions embedded in reproductive health and rights discourse—particularly the concept of individual ownership and control over bodies.

At the end of negotiations at the Cairo ICPD, delegates produced a policy document, the Program of Action (also known as the Cairo Consensus), that was signed by 179 countries. It declared reproductive rights to be universal and asked all nations of the world to foreground women's empowerment as a central component of population and development programs. It also enshrined principles such as advancing gender equality, elimination of male violence against women, prioritizing women's ability to control their own fertility, and the abandonment of demographic targets and quotas.[22]

Despite the efforts of feminist activists, the Cairo Consensus articulated an individualistic, responsibility-centered model for managing fertility in the service of sustainable development goals. While it states clearly that "any form of coercion has no part to play," it nonetheless offers a clear message that successful population programs are predicated on the responsibility of individuals: "The success of population education and family-planning programs in a variety of settings demonstrates that informed *individuals* everywhere can and will *act responsibly* in the light of their own needs and those of their families and communities"[23] (emphasis added). The document also argues that governments should formulate population policies and programs that integrate demographic data into environmental assessments and planning, promoting sustainable resource management and reducing unsustainable patterns of production and consumption.

Although the Cairo Consensus has been widely heralded for enshrining women's SRHR and empowerment and breaking away from demographically-driven population policies, the development of the document was the result of a complicated and problematic process. The ICPD negotiations that produced the document were racked by fractures, bitter struggles, philosophical and political debates, and ultimately a struggle over the future direction of global population policies, including the role of women's reproductive health and rights. Ultimately, the Cairo Consensus united the perspectives of neo-Malthusians, environmental activists, and mainstream and more radical feminists in a pragmatic policy compromise.[24] Many radical and Southern feminists felt that their aims were not satisfied in the consensus document, arguing that it enshrined a weak stance on abortion and watered down language on women's empowerment.

American population-environment advocates also had mixed reactions to the consensus. Some felt that their priorities were effectively silenced in the new focus on individual women and SRHR, while others suggested that the new approach offered natural benefits to the environment and a "win-win" perspective.[25] Over time, many leaders of environmental organizations came to see the population-environment connection as one that could no longer be talked about openly. The post-ICPD euphoria that did exist among some populationist supporters began to wane as funding support decreased and environmentalist interest in family planning weakened in the shadow of the new focus on voluntarism and women's reproductive rights.[26]

Donors and Funding: Navigating the Tides

While they are not the public face of population advocacy, donors have arguably the most important role behind the scenes. Foreign assistance for family planning and other development programs is funded on the basis of key program sectors—global health, environment, democracy and governance, etc.—that, while clearly impacting each other in life, do not often cross funding streams. Providers of this money (governments, private foundations, individuals, the corporate sector) transfer funds to institutions (bilateral agencies, NGOs, intergovernmental organizations [IGOs], and private foundations) to manage and distribute funds to the organizations (multilateral agencies, NGOs, private sector, community-based organizations) on the ground that spend the money on programs and services.[27] Within sectors—global health, for example—donors allocate funds for different programs (family planning/reproductive health, HIV/AIDS, non-transmissible diseases, nutrition, malaria, etc.). Depending on the funding source, the intra-sector allocation for a particular area of development work can fluctuate significantly, making an issue that was once well resourced suddenly impoverished, sending its funders scrambling for resources.

This has been the case for American government-led funding for international family planning, which has been stagnant for decades, after reaching its zenith in the mid-1990s. For years, U.S. government-led foreign aid has been subject to a "global crisis model" in which issues deemed crises mobilize significant attention, and garner disproportionate resources relative to other issue areas.[28] This happened within global health; when the immediacy of crisis thinking about "overpopulation" dissipated after the Cairo conference, attention and funding priorities turned to global AIDS prevention and treatment. Many in the population sector have been seeking to restore population to its former funding prominence within global health ever since. Several former donors from the Office of Population, Health and Nutrition at the U.S. Agency for International Development wrote a 2008 report advocating for more public resources to be allocated for international family planning. Describing an "enormous pent-up and growing unmet need for family planning," the report cites "mistaken" perceptions of declining global population growth, diversion of funding to HIV/AIDS programs, and a general lack of awareness of the role of family planning in economic development as reasons for funding declines. As of fiscal year 2017, U.S. government allocations for international family planning were at $607 million,[29] a figure that has fluctuated, but largely stagnated, over the years since 1994.

While the report does not mention the Cairo Consensus (and the resulting turn toward women's health, rights, and empowerment) as a reason for funding declines, some donors have privately stated that the conference agreement was a cause of funding cuts to international family planning budgets. In an interview, one former USAID population program donor stated, "There's been a decline in family planning funds since the Cairo conference. What happened at Cairo was that there was a focus on women's welfare and reproductive health and less emphasis on population-development, and population-environment. For better or for worse, I think the link between population and environment . . . well, there was the concern that if you were concerned about numbers it would lead to coercive programs, so it became politically incorrect to focus on numbers. The women in the forefront didn't intend to take money away from family planning, but that's what happened."[30]

Given that, in previous decades, population and family planning was one of the most-funded sectors of U.S. development aid, these shifts are striking and deeply disturbing to members of the population-development community. The state of institutional uncertainty about the future of program funding fuels a sense of urgency among those who see new advocacy strategies and narratives as crucial to protecting the sector in which they have made their careers.

One donor I interviewed at a private foundation noted that the decreased funding landscape heightens the need for private donors, who have more freedom and flexibility with how they allocate resources, to be creative in their practices, including directly reaching out to potential grantees (see chapter 3). Yet this process is not without its frustrations: "Some funders offer a lot of money to nonprofits and ask that they do this work. Nonprofit organizations take the idea of connections between population and environment, but don't really get behind it. This rarely works. The ownership is still with the funder." Some donors see the emphasis on Cairo-led language, specifically emphasizing women's reproductive rights and empowerment, as increasingly necessary to avoid controversy and appeal to a wider audience, even though the process often feels experimental and uncertain. As the same private donor lamented, "We're still in kindergarten when it comes to finding the connections, frameworks, models, and prototypes that might actually move things forward."[31]

The donors I spoke to agreed on one strategy in particular: strengthening the scientific base linking population growth to climate change and other environmental problems. During interviews, donors and grantees often spoke appreciatively about the role of science in minimizing controversy, arguing that if the broader public, and particularly legislators, were more scientifically literate, they would no longer see population as a controversial issue. One former USAID donor spoke passionately on the subject: "The general scientific illiteracy of the American people is important. We have climate change deniers who get as much credibility as climate scientists. So I think the weakness of our system to use science to set policy . . . an overarching issue is emotion and politics more so than clear thinking based on our best science." Despite the impulse to separate science from politics, such a teasing apart is nearly impossible. Political goals are precisely what motivate some of the scientific research linking population growth to climate change and other environmental problems. Some private donors, based on their politics, see themselves as uniquely positioned to increase the scientific basis to inform global population policies, and they have pursued grantees accordingly. Chapter 3 delves further into this donor-led advocacy. In the meantime, in what follows, I turn to investigating one of the most enduring and effective ways of engaging public interest in population: the urgency of impending environmental crisis.

The Return of Apocalypse

Climate change comprises a broad range of physical, atmospheric, and material changes to the earth and the atmosphere. It also operates as a set of discourses signifying ideas about the future, resource use, "the global," and personal responsibility, often through the lens of impending crisis. These futures are fundamentally dystopian, characterized by dramatic sea level rise, disappearing coastlines, intense heat waves, and extreme storms. Yet, the future is also already here. Climate change is already disrupting ecological systems and human societies; it is already killing people via heat waves, storms and other disasters, nutritional deficiencies, the intensified spread of infectious diseases, and even chronic disease. The effects are so significant that in 2009 a group of global health researchers declared climate change the biggest threat to human health in the twenty-first century.[32]

Against this backdrop, Bill McKibben published a widely read 2012 *Rolling Stone* article that described the "peril that human civilization is in," arguing that "we're losing the fight, badly and quickly."[33] By crunching a few numbers, he layed out the predicament: human beings are pouring far more gigatons of carbon and other greenhouse gases into the atmosphere, far more rapidly, than expected. In order to remain under the two degrees Celsius warming limit agreed to by world leaders, we would need to add no more than 565 more gigatons of carbon dioxide to the atmosphere by midcentury. As of the date of McKibben's article, we were on track to add more than 2,700 additional tons. Several years later, he followed up with a second, markedly more aggressive message. This time, he likened climate change to war, arguing that "World War III is well and truly underway. And we are losing . . . by most of the ways we measure wars, climate change is the real deal: carbon and methane are seizing physical territory, sowing havoc and panic, racking up casualties, and even destabilizing governments . . . It is not that global warming is *like* a world war. It *is* a world war."[34] Even more terrifying, the war he described was not a war of winners and losers, but rather a zero sum game in which we will inevitably be decimated. McKibben's prescription for a response? Meet fire with fire, declaring war on climate change and mobilizing the kind of response marshalled during World War II.

Climate change discourses are often communicated in this apocalyptic, doom-and-gloom language. They are organized around the idea of "planetary emergency," and if those listening do not retreat into denialism, the focus on crisis is a powerful strategy; it can "mobilize powerful actors around the threat of massive risks and uncertainties."[35] It can also open up the possibility of solutions that are otherwise unacceptable, including population control.

In 2016, a group of bioethicists published an article identifying climate change as one of the most important moral problems of the day, due to its urgency and widespread harmful impacts. Their solution? Population engineering—defined as intentional manipulation of population size and structure—"a practical and morally justifiable means to help ameliorate the threat of climate change."[36] The basis of their argument is the following: population growth and climate change are both proceeding in ways that have destructive impacts on the availability and quality of natural resources, and current international climate mitigation policies and technologies are insufficient to prevent the kind of CO_2 emissions that lead to rapid warming. Planning for climate change mitigation while remaining at current population growth rates is both impossible and morally indefensible, given the global spread and the depth of the impacts global warming is having on the planet. Thus, some level of population engineering is required to address the urgency of the climate problem.

Here is where their argument becomes dangerous: they state that, due to the urgency of the climate problem, the best interventions are neither those that are the least coercive, nor the most, but those *somewhere in the middle*. Achieving quick, climate-stabilizing results morally justifies actions that are along a coercion continuum; after all, they remind us, "We don't have the luxury of solving this problem at a leisurely pace."[37] Specifically, the authors advocate a "global population engineering program" comprised of an expansion of choice-enhancing interventions (such as women's education and comprehensive reproductive health care), along with "preference adjustment"—changing cultural norms and influencing childbearing preferences—as well as "incentivization"—altering the costs and benefits associated with particular reproductive behaviors.[38] Ironically, they conclude with the statement that reducing childbearing is a more attractive climate change mitigation strategy than reducing resource consumption because it is easier, thus helping individuals avoid making unnecessary personal sacrifices.

The article encapsulates many of the arguments writers and scholars are making today, namely that population growth is an important cause of climate change, that slowing population growth will be an effective means of slowing GGEs, and that the urgency of the global climate problem morally justifies—even requires—a narrow focus on fertility reduction. Where it departs from others is in the authors' willingness to consider coercion as morally justifiable in the service of the greater good. However, this is not an unfamiliar argument within environmental circles. In the 1960s and 1970s, in the heyday of public advocacy for population control, environmental scientists openly called for coercive measures, including punitive taxes and other population control policies. More recently, drawing on militaristic language similar to McKibben's, members of the defense community have increasingly described climate change as a national security threat, and talk of impending "climate wars."[39]

The narrative of apocalypse is frightening, not just in what it describes but what it makes possible, including military maneuvers, increasing surveillance and policing of borders, and racial profiling of migrants and refugees. This is linked to another development-industry population narrative: the youth bulge theory. Youth bulge theory says that in populations with large proportions of youth, particularly adolescents and young adults, access to employment and social services is limited, fueling frustration among disaffected young people—rendering youth susceptible to recruitment in rebel or terrorist organizations, and making the societies they live in more vulnerable to social unrest and violence. The theory describes populations with age structures that skew heavily toward youth, with young people in their twenties or younger comprising more than a fifth of the overall population.

Of course, the vast majority of countries with these age structures are in the Global South. The theory is meant to be frightening, and to invoke fear of the "other." A *Newsday* article describes it thus: "Dangerous demographic trends typified by a massive youth 'bulge'—an extraordinary high proportion of young people among the population—all but guarantee increased social instability that few regimes will be able to withstand."[40] Similar discourses had been circulated twenty years prior, in a National Security Council report in 1974 describing "young people, who are in much higher proportions in many less developed countries, [who] are likely to be more volatile, unstable, prone to extremes, alienation and violence than an older population."[41]

Population researchers and NGO program managers have been paying more and more attention to the youth bulge theory in the context of recent international focus on two issues: terrorism and climate change. Take for example a 2010 report by Population Action International on youth bulge theory, *The Shape of Things to Come: Why Age Structure Matters to a Safer, More Equitable World.* The report argues that "the threats to the well-being and security of our world—from HIV/AIDS and terrorism to climate change and poverty—require a bold mix of interventions and partnerships that combine elements of both 'hard' and 'soft' power."[42] The report describes terrorism, security concerns, and fragile or failing states as resulting from youth-heavy population age structures, adding that age structures can and should be shaped through "policies that affect the demographic forces (i.e. births, deaths, and migration) that determine these age structures"—namely policies that support family planning.[43] This approach to "strategic demography"[44] was replicated just two years later, when a group of researchers at the University of California, Berkeley convened a meeting to discuss how to prevent "a huge humanitarian catastrophe" from occurring in the Horn of Africa. Over a hundred people gathered to review population projections and models of climate-driven food and water shortages, and to pore over data arguing that rapid population growth, extreme temperatures, and expected drought would likely create conditions leading to "the largest involuntary migration in history."[45] The solutions proposed at the meeting focused on agricultural adaptation, water management, and predictably, voluntary family planning for women delivered through a human rights approach, arguing that these interventions must be "immediate and on a large scale."[46]

These discourses make it clear that members of the population network are nimble in applying the Cairo Consensus approach to family planning to a broad range of conditions, and in doing so, they harness the exact kinds of discourses Cairo activists worked so hard to reject. Operating within a discourse of entrenched gender stereotypes of "dangerous" Southern men, they harness a powerful set of images that have already taken root in U.S. policymaking groups and in the national psyche. In an era when terrorist threats are perceived around every corner, when terrorist attacks are quickly used to racially profile young men from the Middle East and Africa, and when this generalized perception of Muslim-led terror becomes embedded in U.S. policy in the form of travel bans and refugee exclusion, youth bulges operate as an explanatory framework for racialized framings of population problems. They also make a case for the familiar technological quick-fix: contraceptives.

Conclusion

This chapter demonstrates how population narratives and discourses, far from silent or invisible, are continually revived through popular media, development interventions, and various forms of policy advocacy. Climate change, however, is making these narratives more salient, visible, and urgent in an era of heightened environmental and social risk. Characterizing population through crisis narratives has a long history among neo-Malthusians, particularly those tasked with developing scientific theories that link population to environmental problems and policy solutions. Chapter 2 delves into these questions, investigating the ways key actors constructed population as a scientific and environmental problem across the twentieth century—and the roles of politics and policy agendas in the process.

2

How Population Became an Environmental Problem

One afternoon in 2009, a friend and I flew home to Northern California from our Costa Rica vacation, gazing through the window and marveling at how many trees covered the land just south of San Francisco. "It's so beautiful," he said. "People should start using some birth control so that we can get this back in San Francisco!" A lifelong resident of the city, he was struck by the contrast between the view from the airplane and the landscape of concrete, steel, and glass that had shaped everyday life in his hometown. I, on the other hand, was surprised by the casual way he linked deforestation and population growth without any broader context. Those details didn't matter. The paradigm he was operating in said that a lack of forest cover had to be the result of too many people in the region. When I asked him where he thought his idea originated, he responded defensively: "What do you mean? It's obvious, it's just common sense. My idea came from me. No one has to tell me there are too many people around. You can look out the window and see that."

I tried to probe further, suggesting that he had learned this idea from somewhere—a class, a textbook, newspaper article, television ad, or conversation with an environmentalist friend—but the conversation quickly degenerated as his insistence on the obviousness and inevitability of population growth's environmental destruction met with my own unyielding position that his ideas did not originate in his own mind. Things became heated. As we descended, both toward the airport, and more deeply into our entrenched positions, I was struck by how the conversation brought to life a key issue I'd been grappling with in my research: how and why populationist approaches are accepted, and stridently defended, as common sense. It seems to be a natural and obvious connection: the more people there are consuming natural resources, the more those resources will inevitably be depleted and become scarce. Of course, the real picture is far more complex than that. And our ideas of what counts as common sense, particularly in the context of population-environment linkages, are heavily shaped by the assertions of scientists and environmentalists, many of whose ideas were historically shaped by social and political concerns.

As this chapter will demonstrate, these concerns lay at the foundation of much of the thinking about carrying capacity and planetary limits. This is not to say that there are no limits to the earth's ability to sustain and renew itself and its resources, nor am I arguing that human numbers do not play a role in stretching those limits. However, what I do argue is that there is not, and never has been, a single, evidence-based model that has successfully calculated or predicted the global environmental impact of human numbers *alone*. Local context, resource consumption, polluting technologies, state- and corporate-based resource extraction and pollution, and the environmental impacts of military operations all make it impossible to produce such a number on a global scale.

Population has not always been thought of as a problem, and certainly not an environmental one. Rather, it has been socially constructed over time as a problem through the efforts of people who were deeply concerned with managing a range of issues: "socially undesirable" traits like criminal tendencies, alcoholism, mental illnesses, cognitive delays, unwed pregnancies, and dependence on welfare; global geopolitical instability and armed conflict; immigration; economic development in newly independent former colonies; and the balance between human consumption of natural resources and the earth's capacity to regenerate those resources. A closely related set of concerns focused on the roles of technology and modernization in addressing these questions. The taken-for-granted idea that population growth is a threat to nature and the environment does not in fact reflect an essential, immutable biological reality. Instead it reflects long-standing debates among scientists, activists, academics, and policymakers working to define population problems, their impacts, and how to solve them.

What those problems are, and the strategies designed to solve them, have changed significantly over time. For example, at the start of the twentieth century, scientists and politicians—including environmentalists—were quite concerned about population *decline*, and were focused on increasing the population of the white, U.S.-born middle class. As this example suggests, while population itself has long been a fraught topic in the United States, there was not a singular or essential twentieth-century "population problem," but instead a set of diverse and sometimes competing ideas making up a field of population problems. These problems were at the heart of complicated struggles and debates that engaged questions of policymaking, resource use, ecological stability, international relations, geopolitical stability, race, and gender. Throughout these debates, struggles have centered on women: whether and which women should have more children or fewer, and how to bring that about through clinical interventions, laws, social policies, or moral exhortations. These are also racialized struggles. In the U.S., whiteness in particular has been central to populationist ideas about natural resources, national identity, and global geopolitical security.

This chapter is about the twentieth-century history of population science. It offers a critical feminist reading of the scientific ideas, theories, and concepts that have produced knowledge of population as an object of inquiry, building it over time as a problem of nature and the environment. Traversing the late 1700s to the 1970s, when the core theories of population science were produced, the chapter explores several questions: how did population become an environmental problem? What are the relationships between science, politics, and policy concerns in population-environment arguments? How did the logic of population problems come to be known as common sense? And a central, related question: how do race, gender, and class inform these debates? I trace these questions through a primary focus on three conceptual/paradigmatic and historical moments: the development of eugenic theories and nationalist discourses in the 1900s–1920s; modernization and demographic transition theory in the 1930s–1950s; and the concept of carrying capacity as developed and disseminated by environmentalists in the 1940s through the 1970s.

Defining the Problem: The Emergence of Population

From the earliest days of population being perceived as a bounded subject and area of inquiry, it was situated within the context of political concerns, namely questions of state and territory governance, international security, and the need for an optimal balance of land-based resources to meet the basic needs of individual members of society. These questions emerged in Europe in the sixteenth and seventeenth centuries as part of efforts to identify the basic blueprint of the ideal state.[1] The question of ideal population size was the subject of intense debate, reflecting contrasting views across Europe. Fears of overpopulation and the ill effects of high population density began to flourish in England and Germany, while writers in Italy and France argued in favor of abundant populations as a source of political stability.

Views on the significance of large populations varied widely among European thinkers; they were characterized both as providing benefits for state strength, sovereignty, and security, as well as draining resources and driving state decline. A broad change in the approach to population began to develop in England as a result of the rather sudden growth of the poor—a transformation that arose from both an absolute growth in human numbers, as well as sudden increases in capitalist-driven inflation and enclosures of previously common agricultural lands, which left farmers devastated.[2] Hunger became a state of daily life for the newly poor, a class that swelled dramatically, seemingly overnight. It was in this context that British state authorities began to identify poverty, crime, and "overpopulation" as the prevailing national problems of the day, leading to the institutionalization of the Elizabethan Poor Law in 1601. The law provided money, food, and clothing to the settled poor, who were temporarily out of work. Amid heightened population anxiety in England, European writers elsewhere continued to argue that a robust and sizable population was the source and symbol of strength of a country's international leadership. Contrasting views dominated population thinking throughout the seventeenth century: while some authors associated large populations with idleness, waste, poverty, and potential criminal tendencies, others argued that large populations were a necessary source of labor for state projects.[3]

In the eighteenth century, a total revolution in population thinking occurred with the appearance of Malthusianism. British cleric and political economist Thomas Robert Malthus wrote the first book of his six-draft *Essay on the Principle of Population* in 1798, theorizing that there were two basic forces that operated to create and sustain life: human sexuality, or "the passion between the sexes"—a necessary and vital component of human nature—and food production ("production in the earth").[4] These two forces were unequal, and subject to the laws of nature, which were fixed, predictable, and determined by God. In this schema, population grows exponentially while food production proceeds at a much slower pace; as a result, the great law of nature was required to impose natural limits on population growth. These limits, or checks, included misery, which he defined as starvation, disease, and death. Without this fixed law of nature, unchecked human population growth would inevitably exceed the earth's capacity to produce adequate food for human survival. Malthus put forth these arguments as fundamental laws of a nature that binds all living things (plants, animals, and humans) together in the same set of natural limits. In his framework, famine is nature's resource, causing premature death to intervene when the power of population surpasses that of subsistence.

Although Malthus was not the first to make these arguments, he was the first to synthesize political and scientific arguments about population into one coherent theory. In addition to making arguments about the laws of nature, he argued that state welfare programs like the British Poor Laws prevented the poor from adopting the responsibility required for a strong work ethic and artificially prevented what nature intended: human starvation. He claimed that the Poor Laws directly facilitated population growth by removing nature's limits—misery, starvation, and death—thereby inciting the poor to reproduce. Natural limits were also intended to impose moral boundaries on human kind: man's highest moral and social potential could only be achieved through hard work and responsibility, which state intervention made impossible. In other words, social policies improving the welfare of the poor prevented nature from keeping human growth and food resources in balance, and kept the poor from contributing effectively to society through hard work.

Malthusian theory was groundbreaking. Population was no longer an arena of abstract theorizing; it was now an object of quantitative inquiry. Future population growth, and its effect on food resources, could be projected by analyzing and developing theories about current population trends. Malthusian theory also irrevocably influenced the development of scientific thinking, directly shaping the expansion of scientific theories and methods in the fields of eugenics, demography, and the ecological theory of carrying capacity.

Operationalizing Malthusianism: Eugenics, Conservation, and the New Nationalism

Malthusian ideas had a direct impact on the development of genetic science through the controversial field of eugenics. Charles Darwin was heavily influenced by Malthus as he developed his theory of natural selection, writing in his autobiography, "I happened to read for amusement Malthus on Population and, being well prepared to appreciate the struggle for existence which everywhere goes on from long-continued observation of the habits of animals and plants, it at once struck me that under these circumstances favourable variations would tend to be preserved and unfavourable ones destroyed. The result of this would be the formation of new species."[5] Several decades after Darwin published *Origin of Species* in 1859, his cousin Sir Francis Galton, a British anthropologist, read the book and became interested in studying human heredity. Based on Darwin's ideas, Galton postulated that "innate moral and intellectual faculties," which are "so closely bound up with the physical ones," could be selectively encouraged through human breeding, thereby producing people of superior stock.[6] In other words, Galton thought that these characteristics were genetically grounded and transmitted from one generation to the next via DNA.

In 1883, Galton coined the term "eugenics," which he drew from the Greek word "eugenes," meaning good or well born, defining it as "the science of improving stock."[7] His theories were grounded in the idea that traits like intellectual and physical prowess, moral fiber, personality, and criminal tendencies were traits received at birth through genetic inheritance. Galton also viewed human traits and characteristics as biologically rooted in race and transmitted through heredity, arguing that the most desirable traits were possessed by races of superior stock—namely, European races. While he communicated his ideas in scientific terms, eugenics was initially conceived as both a scientific and a political enterprise; Galton advocated state policies such as encouraging early marriage between those deemed socially superior, while refusing state welfare to those whose children displayed "inferior qualities."[8]

Galton's eugenic ideas reflected broader concerns about biological and social deterioration, concerns that informed much of the research that was produced over the next century in Europe and North America on questions of race, sexuality, reproduction, nature, and environment. American eugenic ideas were also applied to natural landscapes and conservationist thinking. In California, for example, the landscape was profoundly shaped by eugenicists, many of whom were prominent conservationists. The Save-the-Redwoods League was founded in 1918 by three California men who were key actors in the American eugenics movement: Madison Grant, Henry Fairfield Osborn, and John C. Merriam. Madison Grant wrote *The Passing of the Great Race*, a polemic railing against the rising populations of nonwhites and immigrants, and the impending extinction of the "great race" of white Protestant Americans. The League's board was made up of wealthy white men, many of whom were landscape architects, engineers, and natural scientists. Charles Matthias Goethe, a long-term member and supporter of the League who was hailed by the *San Francisco Chronicle* in 1966 as "America's Grand Old Man of the Conservation Movement" following his death, was a widely renowned conservationist, member of the Sierra Club and Audubon Society, financial supporter of plant biology and genetics research, natural scientist, and eugenicist. Goethe's memoir compared the European American migrants who settled in the West to cacti and other hardy plants that were able to thrive in rugged conditions, describing westward expansion in terms of the natural selection of hardier human "stocks" and elimination of "weaklings."[9] Many of the League's members were either members of national eugenics organizations or explicitly endorsed eugenic immigration policies and ideas of Aryan and Nordic supremacy.

Eugenicists in early-twentieth-century California drew symbolic associations between natural landscapes, redwoods in particular, and the white race. Endangered redwood trees became emblematic of imperiled human genetic stock, specifically that of Anglo-Nordic people, who these eugenicists saw as superior. They entwined narratives of colonial westward expansion and survival on the harsh landscape with the survival of trees in the midst of encroaching human, plant, and genetic threats. In this register, genetic stock became a way of talking about both trees and people in terms of survival, conservation, protectionism, greatness, and the threat of encroachment by inferior stock.[10] People like Osborn and Grant in the 1920s applied eugenic arguments to immigration, casting it as a threat to the integrity of vital plant and animal species, as well as conservation more broadly. Their target was southern Europeans, and their language was every bit as racist as that which would follow decades down the line. In their schema, genetically inferior people threatened to overwhelm native-born whites through immigration, and in tandem, to overrun and overwhelm wilderness. They also saw the immigrant masses as representative of cities encroaching on wilderness areas.[11]

These ideas were front and center in American politics and in the nascent conservation movement in the early twentieth century. President Theodore Roosevelt, a major proponent of eugenics and conservation, convened the first national conservation conference in 1908, bringing together members of Congress, scientists, governors, members of the Supreme Court, journalists, and leaders of environmental organizations. It was an unprecedented meeting. Roosevelt argued against rampant timber logging and deforestation, lamenting the impacts it had on soil erosion and sullied waterways. The two central priority areas of the conference were the need to "keep waterways free for commerce," and to conserve natural resources, both of which he linked to the future of the nation itself. The conference pushed forward the point that "the nation needed to manage its forests with more foresight and wisdom, not just so that future Americans would have trees to use but so that the future of the entire nation would be bright."[12] Conservation, or "wise use" of natural resources, was, in Roosevelt's eyes, "the great material question" of the day.[13]

These ideas were central to the new nationalism of the time, which focused on individual opportunity combined with civic engagement and good citizenship, in protection of what Roosevelt saw as the American ideal: self-advancement and increasing wealth and prosperity for American families working on their own land. Roosevelt's vision, however, was reserved exclusively for white Protestant Americans. He was an ardent eugenicist who saw race and nationalism as deeply intertwined, and he perceived a racial threat in the form of immigrants from east and southern Europe—one that threatened the very existence of white Protestants. In an October 18, 1902 letter, he wrote of "what is fundamentally infinitely more important than any other question in this country—that is, the question of race suicide, complete or partial." The solution he proposed was to build robust families centered on marriage and abundant children, arguing that it amounted to a moral and national duty: "The man or woman who deliberately avoids marriage, and has a heart so cold as to know no passion and a brain so shallow and selfish as to dislike having children, is in effect a criminal against the race, and should be an object of contemptuous abhorrence by all healthy people."[14]

The duty Roosevelt described was primarily a gendered one centered on white women. In subsequent speeches, he outlined women's special embodied role, arguing that they bore a particular responsibility to the nation: that of bearing children to build and strengthen the race. In his speech to the National Congress of Mothers, he outlined these ideas as follows:

> No piled-up wealth, no splendor of material growth, no brilliance of artistic development, will permanently avail any people unless . . . the average woman is a good wife, a good mother, able and willing to perform the first and greatest duty of womanhood, able and willing to bear, and to bring up as they should be brought up, healthy children, sound in body, mind, and character, and numerous enough so that *the race shall increase and not decrease* . . . If the average family in which there are children contained but two children the nation as a whole would decrease in population so rapidly that in two or three generations it would very deservedly be on the point of extinction, so that the people who had acted on this base and selfish doctrine would be giving place to others with braver and more robust ideals. Nor would such a result be in any way regrettable; for a race that practised such doctrine—that is, a race that practised *race suicide*—would thereby conclusively show that it was unfit to exist, and that it had better give place to people who had not forgotten the primary laws of their being.[15] (emphasis added)

Roosevelt's racial ideas were deeply entwined with his conservationist values. He established the United States Forest Service in 1905, and used the American Antiquities Act of 1906 to establish national forests, bird reserves, game preserves, national parks, and other monuments. But these resources were created for the enjoyment of white Protestant Americans. Dispossessing Native Americans of their lands—known as "Indian Removal" at the time—was central to the project of designating national parks and other wilderness areas as recreation sites for whites. National parks were living Edens, places to experience sublime nature in the form of beautiful scenery, animals, trees, flowers, mountains, lakes, and streams. Designated wilderness areas were carefully managed to remove encounters with "wild" Indians. Early environmentalists like John Muir, a preservationist and founder of the Sierra Club who was a close friend to Roosevelt, saw the project of removing and resettling Native Americans onto reservations as necessary to the conservation of wilderness. Muir, who described Native Americans as dirty, gross, grim, and degraded in his travel journals, contrasted their "uncleanliness" with the pristine beauty of nature.[16]

Conservationists promoted eugenics as vital to the optimization of both nature and humans. In a 1909 report for the National Conservation Commission (appointed by Roosevelt), Yale economist Irving Fisher wrote that "there is every reason to believe that human beings are as amenable to cultivation as other animals and plants," and concluded that "the problem of the conservation of our natural resources is therefore not a series of independent problems, but a coherent all-embracing whole. If our nation cares to make any provision for its grandchildren and its grandchildren's grandchildren, this provision must include conservation in all its branches—but above all, the conservation of the racial stock itself."[17] At the conclusion of his report, Fisher made ten comprehensive recommendations to promote this conservation, the final one of which was eugenics, which he defined as "hygiene for future generations," in order to prevent marriage between the "unfit," as well as ensuring the "unsexing," or compulsory sterilization, of rapists, criminals, "idiots," and "degenerates generally."[18]

Despite its early associations with scientific racism, eugenics was actually a somewhat diverse field, divided into two main areas of inquiry: positive eugenics—encouraging the childbearing of those deemed to be of better stock—and negative eugenics, which focused on strategies to limit the procreation of the unfit. As Alexandra Minna Stern's study of twentieth-century American eugenics argues, prevailing narratives of eugenics today often misunderstand, and therefore misrepresent, the field's complexity and its legacy, assuming that eugenics disappeared in the U.S. after Nazism and WWII due to its association with scientific racism. This was far from the case. Eugenics researchers simply redirected their efforts into new research agendas, moving in two key directions. The first direction primarily focused on individual genetics counseling for couples hoping to have children. The second was that of international population control, where eugenic ideas about improved population quality were integrated with modernization theory and development interventions in international family planning.[19]

Making Modern Subjects: Demographic Science and Development

Greene's study of the role of Malthusianism in shaping U.S. domestic and foreign population interventions describes population as a governing apparatus, functioning to "invent, circulate, and regulate public problems."[20] This apparatus makes the "population crisis" intelligible, so that it is possible to define and solve population problems. It also makes modern subjects possible through fertility regulation. Malthusianism translates reproducing bodies and populations into problems that can be solved through modernizing tools and technologies, including demographic and other scientific study, modernizing discourses, and interventions such as family planning and contraceptive use. In creating his theories, Malthus "helped make visible the reproductive practices of the working classes as an object open to change. In other words, he pulled procreation out of the realm of the natural and into the realm of the governable."[21] This cannot be overstated: what was previously understood to be natural—procreation—was now reframed as something that could be artificially regulated through study and intervention.

Malthusian ideas became grounded in the quantitative sciences through demography, a field that emerged as the "offspring of mixed parentage and stormy unions" between activists, ideologues, biologists, and policymakers, among others.[22] The first meeting of what would eventually become the Population Association of America (PAA), the largest society of demographers in the U.S., convened in 1930. Birth control advocates, eugenicists, immigration control advocates, a population biologist, agricultural economists, and statistics gatherers from government, industry, and academia came together, the activists far outnumbering the scientists, to decide how population studies would be defined, and how population knowledge would be applied to social concerns.

At the time, "biological Malthusianism," which incorporated social Darwinism and racist ideas, dominated the discussions. This new approach recast the smaller families of the middle and upper classes—which Malthus had lauded as economically beneficial for the state—as dangerous, a sign of "selfishness" and "excessive materialism."[23] Biological Malthusians also saw feminism as part of the problem, as most of the adherents to the birth control movement at the time were educated, U.S.-born, white women, precisely those who population scientists wanted to see reproducing. There were divergent goals among the group: birth controllers concerned about freedom for white women and improved living standards saw declining populations among native-born whites as a good sign, whereas biological Malthusians were pessimistic and sought to intervene and increase birth rates of native-born whites. A third, growing group of "population scientists" emphasized compiling population statistics to make questions of population change more empirical in basis.

These statisticians were the earliest demographers, along with biologists, eugenicists, economists, and sociologists, and their work was directed toward institutionalizing demography as an academic field of study through universities in the 1920s and 1930s. Although demography became an institutionalized science through the academy, quantitative analysts did not reject more ideological positioning. As Hodgson notes, the "work of these empirically minded population scientists is distinguishable from those who wrote on population themes as movement advocates . . . [However,] belief in the empiricist tradition did not preclude ideological commitment. During the first third of this century, population scientists largely accepted the precepts of biological Malthusianism."[24] For example, when the PAA was formed in 1931, its first president was Henry Fairchild, a "nativist with clear eugenicist leanings who had an academic post teaching courses in population studies."[25]

A central figure in demography was Frank Notestein, who articulated the theory of demographic transition in 1945. The theory outlined a normative four-stage model of demographic transition from high-fertility, high-mortality population trends, to a pattern characterized by low fertility and low mortality, based on historical population trends that began in Europe in the eighteenth and nineteenth centuries. In the first stage of the transition, mortality is decreased via improvements in agriculture, industrial production, and improving human health and life expectancy. Notestein summed up these innovations up as encompassing modernization. Stating that more than half the world had not begun the demographic transition, he argued that this transition could be brought on through Western intervention, such that Global South countries could effectively follow the European model. Notestein's concept of demographic transition was grounded in capitalist development, based on the logic that only a significant increase in economic production could bring the improvements in social conditions, health status, and overall social welfare necessary to reduce fertility for the world's poor. These concerns had by this time become a core component of U.S.-led international development projects, which dovetailed over time with neo-Malthusian population arguments.[26]

Demographic transition has alternately been described as a theory, a historical model, a predictive model, and a descriptive term. It is also a tool of policy-oriented science. As Szreter notes, "The idea of demographic transition was itself the product of a particular conception of social science as a guide for policy, a science employing a positivistic methodology that was simultaneously investigative and predictive."[27] Moreover, demographic transition was predicated on modernization theory: "The theory held that fertility would only fall as a result of the cumulative mutually reinforcing spectrum of effects consequent on full-scale industrialization and modernization: enhanced survival; a growing culture of individualism; rising consumer aspirations; emergence of huge and socially mobile urban populations; loss of various functions of the family to the factory and the school; and decline of fatalistic in favor of conative habits of thought."[28]

Scholars have discovered abundant historical evidence to refute the empirical basis and analytic value of the demographic transition; however, the concept continues to dominate both demography and international family planning policy and programming. Demographic transition was developed to describe a historically contingent set of events, not a stable, static, or universal set of trends. It has endured in part because of its ideological pull: by the early 1960s, "many demographers, including many illustrious figures of the discipline, decided that their task was not just to interpret the world *but to change it*"[29] (emphasis added).

Demographic transition theory is based on a set of observations of fertility and mortality change over time across Europe over a period of more than two hundred years. Early supporters of the theory did not presume that these shifts were uniform across the region, nor that the rate at which these shifts occur could be the same across regions, given that they were precipitated by urbanization, delayed marriage, improved sanitation and hygiene systems, modern health practices, and disease control. However, there was general pressure at the time for a policy solution to rapid population growth in the Third World. Birth control advocates had long argued that modern birth control methods could be used to intervene on high fertility rates; social scientists increasingly began to adopt this view as well. By the end of the 1950s, demographers were advocating the position that individual changes in attitude and behavior could shift the broader demographic landscape—a reversal of the majority position from a decade earlier.[30]

As an applied science, demography was designed to influence policy. These policy linkages became formalized in the 1940s and 1950s through a dramatic increase in network-building between government economic planners, foreign policy experts, professional demographers, corporate leaders, and directors of philanthropic organizations, who shared the goal of "re-working . . . demographic knowledge to make it more 'user friendly' to policymakers."[31] To address this concern, they adopted a three-pronged approach to build and widely disseminate demographic knowledge: data-collecting missions to Global South countries; institutionalizing professional demography as academic and policy science via university-based population centers; and influencing professional demographers to shift their academic paradigm, the demographic transition theory, to better support prevailing political goals of the day.

At the urging of John D. Rockefeller III, who had long been interested in funding contraceptive programs, the Rockefeller Foundation funded a research mission to Asia after World War II to study the demographic context in Japan, Korea, Indonesia, and Taiwan. The delegation included Fred Notestein and Irene Taeuber, prominent American demographers at the Princeton University Office of Population Research, who produced a report making recommendations for foundation support for population interventions in the area. Despite its scholarly tone, the conclusions were clear: Asia's growing populations presented an imminent crisis for the region. The document was distributed beyond the foundation, to academics and policymakers, other foundations, and military officials.

This pilot set the blueprint for future demographic missions to the Third World throughout the 1950s, in which delegations led by prominent demographers would produce reports for the Washington foreign policy community. Through these missions, demographic knowledge became increasingly international, and more closely and formally linked to U.S. foreign policy.[32]

Demography became institutionalized in the U.S. at university research centers, with the support of the Ford Foundation and Population Council. Funders also provided support to train elite foreign scholars in demography at American universities, who would then return to their countries and work in demographic research centers or government agencies. Thus, American demographic theory and methods were spread around the world, facilitated by private philanthropic organizations. This model continued through the 1960s, with the Ford Foundation and Rockefeller funding policy-directed independent policy research centers and other scientific programs.

Demographic transition theory and the quantitative demographic studies produced on the basis of the theory were also central to race-making projects. Studies conducted in Puerto Rico were utilized to define reproductive difference in ways that made colonization both possible and necessary, in order to modernize families. Targeted interventions into venereal disease, prostitution, and "immoral sexual relations" became central to the U.S. colonial project there via public health programs.[33] By constructing sexual and reproductive differences as pathological, this work also served to construct and solidify ideas of racial difference.

Further, McCann has argued that demographic analyses contributed to hegemonic gender norms in the twentieth century. For example, demographic quantification focuses on aggregation, an approach that takes the so-called average man or woman in a society as its main unit of analysis. But the ideal subjects put forth in this model were fictional: men who based their reproductive decisions on market logic and economic self-assessments, and women who only entered the picture as dependent wives, despite the fact that their bodies were the terrain on which measurements of birth rates, contraceptive use, and marriage rates were made. This construction of average, ideal men and women, operationalized through demographic transition and deployed through development interventions, served to homogenize a national "us" and highlight contrasts with (as well as make for easy comparisons to) racialized others whose reproductive, marriage, and kinship practices did not fit the model.[34]

Much of the demand for demographic models and statistical analyses has come from institutions with action agendas, including governments, foundations, and international agencies. To receive funding and the basic data to analyze, demographers have had to tailor their work to be policy-relevant and responsive to the applied interests of their clients. Thus, the field has continuously had to contend with two central dilemmas: first, making and doing science, in order to establish itself as a scientific field separate from policy and politics; and working closely with politically oriented actors. These practice-based realities have created a context in which, as Greenhalgh notes, the field has at times had to develop and promote theories created to meet the standards of its supporters—even when this conflicts with scholarly standards of empirical accuracy.[35]

The Roots of Carrying Capacity

At the same time that the field of demography was being established, biological and ecological scientists were hard at work on their own search for quantitative explanations of the world. The history of carrying capacity and its application to the population problem is one of a search for numbers: a number, calculation, or equation that can quantify the correct number of people to limit population to, in order to achieve balance with natural resources and prevent or reduce environmental degradation. Such a number has never been found; however, the quest to find it has never waned. The reasons for this are complex. The language of mathematics, which is highly structured and rule-bound, projects a sense of uniformity and rigor. This language is easily transportable across vast distances of geography and culture, minimizing the need for intimate knowledge of locality and community. Moreover, the discourse of mathematics helps to "produce knowledge independent of the particular people who make it."[36] Quantification, and its close identification with scientific objectivity, is often taken to signify a "set of strategies for dealing with distance and distrust," a universalizing technology for representing truths in ways that minimize complexity and erase the position of the person producing the knowledge.[37]

Demography lent strong tools to this quest through projections, models, and statistical methods. As the previous section demonstrated, both demography and eugenics also produced ideas of the centrality of racial difference relative to population size, growth, and reproductive behavior across racial, cultural, and geographic lines. Environmentalists would carry this notion further, embedding it in arguments about earthly limits and the need for a raced and classed lens for population interventions. Unlike interventions developed to address public health, carrying capacity arguments would be used as a basis for arguing against the optimization of human health, and for a cold calculus in determining the relative value of lives—particularly those deemed environmentally destructive.

Population biology and population ecology have relied heavily on the logistic growth curve, a model developed by American zoologist Raymond Pearl. Working with statistician Lowell Reed and heavily influenced by the ideas of mathematician Pierre Verhulst, Pearl was the central figure responsible for circulating the idea of the logistical growth curve as a natural law of population growth. Pearl was an editor of the *Journal of Heredity*, where he oversaw the publication of articles arguing that the communities with the highest birth rates were the least wealthy and educated, and that "the old stock had been overrun."[38] He studied yeast and fruit flies bred in glass jars, extrapolating from these results to argue that human population growth follows a strikingly similar pattern to that of other organisms in nature, one that fits into a smooth "S" pattern known as the logistic curve.[39]

In this model, population grows exponentially until reaching a limit defined by the environment, at which time it begins to decelerate and decline. Pearl argued that this process is biological, asserting that "all the complexities of human behavior, social organization, economic structure, and political activity, seem to alter much less than would have been expected the results of the operation of those biological forces which basically determine the course of the growth of populations of men, as well as those of yeast cells and . . . flies."[40] In other words, he assigned biology explanatory power as the determining force in population growth and, ultimately, its limits.

Pearl was the first to take population from an abstract object of concern to a scientific object to be studied and manipulated through direct intervention. Unlike Malthus, who viewed food production and population growth as fixed rates such that population would inevitably overwhelm food production, Pearl saw these rates as adjustable based on interventions through state policies and technologies. He was concerned about global resource limits and was an early critic of human patterns of resource consumption, arguing that "the volume and the surface of the planet on which we live are strictly fixed quantities. This fact sets a limit."[41] However, the logistic growth curve was widely rejected by economists and demographic statisticians at the time that Pearl promoted it. The model did not fit much of the existing data on population trends in many countries around the world, and was thus ineffective in predicting future population trends.[42] Yet, over the next two decades, it became incorporated into measures and formulas tracking population growth in both population biology and population ecology, fields that heavily influenced the development of human demography.

Why did this happen? One explanation is provided through Pearl's personal efforts, including mounting a massive public relations campaign among his natural scientist peers, publishing numerous papers, articles, and books attesting to the validity of the logistic curve model. In one, he laid out three principles outlining what he saw as the biological basis for the behavior of all organisms: the drive for personal survival, the urge to reproduce, and genetic and somatic variability. To them, he added a fourth, determining factor: the environment. "In considering the biology of populations, one aspect of the environment is of particular importance, both theoretical and practical," he argued. "This is available space. The number of organisms in the population taken in relation to available space determines density, a major factor of significance in population biology."[43] Pearl's goal was to produce a model that could make standard predictions about future population growth at regional and global levels. What initially emerged as an intuitive, *a priori* perspective that was initially discredited had become grounded as central to understanding the biological basis of population growth and decline. It was also incorporated decades later into theories of carrying capacity when they were applied to human population growth.[44]

While Pearl was working to promote his theory of the logistic curve, efforts were underway to organize the first international union of population scientists (the International Union for the Scientific Investigation of Population), whose priority concerns were soil productivity and political borders. Early conveners were also concerned with optimal population density, carrying capacity, and migration laws. These were simultaneously scientific questions as well as geopolitical concerns and questions of earthly and spatial phenomena, centered on population size and distribution in relation to land. Malthusians, economists, geographers, and early demographers involved in the union all shared the belief that the question of population density, and the differences of these densities between nations and regions of the world, was central to international relations. A key question of international relations, then, had to be that of how to resolve the problem of uneven population distribution.[45]

Many of these scientists looked to the then-recent history of World War I to argue that growth and uneven distribution of populations was a likely cause of war. As a result, they posited a better distribution of global populations, based on optimum density, as a necessary solution. Bashford's analysis of thousands of publications on world population from the 1920s and 1930s reveals that the most common statement found in these documents was that "overpopulation, understood in terms of comparative density, caused war."[46] Land wasn't the only concern, at least not in itself; the scientists were primarily focused on food production. Their concerns about fertility were just as much about soil fertility as they were about women's reproductive fertility. Many early conservationists and population scientists at the time were first concerned about solving problems of land and food distribution; managing women's fertility was simply the most expedient means to that end.

Carrying capacity has become one of the single most important scientific concepts in neo-Malthusian arguments about population growth and environmental limits. It is centrally concerned with the question of limits—the idea of people around the globe sharing a limited living space characterized by constraint and crowdedness.[47] Ironically, the phrase itself did not originate with Malthus, nor with studies of population. Rather, the term "carrying capacity" was first coined in 1845 as a reference to the tonnage, or storage capacity, of ships, and was mainly invoked in discussions of international trade disputes. It was later applied to animals and rangeland management. Aldo Leopold was a prominent and vocal supporter of this model. In the 1930s, Leopold took up the concept in his analysis of a population crash among Kaibab deer, creating a theory of carrying capacity that would allow game managers to increase or decrease populations of game by manipulating factors like environmental habitats, controlling populations of predators, and relocating animals or releasing the captive bred into the wild. He defined carrying capacity as the number of livestock a given area of land could support without the land degrading. There was significant debate over the concept; many range scientists of the time rejected carrying capacity outright. However, some scientists forged ahead with using the term, and by the 1940s, they applied it to people. In a 1941 speech entitled "Ecology and Politics," Leopold argued that "every environment carries not only characteristic kinds of animals, but characteristic numbers of each . . . that number is the carrying capacity of that land for that species,"[48] and extrapolated these ideas from animals to humans: "Perhaps the present world-revolution is the sign that we have exceeded that limit, or that we have approached it too rapidly."[49]

Carrying capacity later emerged as an expression of state power and control through selective application by state authorities. Because of its development and circulation through scientific discourses, carrying capacity became so embedded in population sciences that "even when carrying capacities proved illusory, they provided an appearance of objectivity, rationality, and precision to policies that might otherwise have been revealed as politically or economically motivated."[50] Following World War II, concerns about population, resources, and international conflict exploded among American researchers and policymakers, refracted through the lens of environmentalism. In 1948, William Vogt published *Road to Survival*, the first tract articulating a global neo-Malthusian carrying capacity: "The lot of each . . . is completely dependent on his or her global environment, and each one of them in greater or less degree influences that environment. One common denominator controls their lives: the ratio between human populations and the supply of natural resources, with which they live, such as soil, water, plants, and animals."[51] He concluded that carrying capacity was a result of the ratio between biotic and environmental factors, an equation that "every minute of every day touch[ed] the life of every man, woman and child on the face of the globe."[52]

While drawing scientific conclusions, Vogt interpreted and communicated them in political terms—the apocalyptic language of war, chaos, and death. He argued that if state leaders ignored the relationships described in his equation, death and destruction were inevitable: "There is little probability that mankind can long escape the searing downpour of war's death from the skies," leading to a state of global chaos in which "at least three-quarters of the human race will be wiped out."[53] His book was written as a polemic, one directed at the poor: in it, he railed against doctors for keeping poor colonial populations alive and able to multiply; he excoriated the poor for reducing and degrading land and soil quality, pushing it to the limits of its productivity; and he advocated harsh policing of the threat posed by growing populations in Asia. Among the solutions he advocated, controlling human population growth through contraception was primary. Vogt recommended that the Food and Agriculture Organization (FAO) integrate population control into its conservation and food production programs, including denying food aid to India and China, since with such aid one would "keep alive ten million Indians and Chinese this year, so that fifty million may die five years hence."[54] Although he argued that contraceptive use should be voluntary, Vogt supported providing financial bonuses to individuals who agreed to be permanently sterilized. *Road to Survival* was a runaway success, and three years later, Vogt was appointed national director of the Planned Parenthood Association of America, a position which he occupied for the next thirty years on an agenda favoring the distribution of cheap contraceptives, strategies to increase contraceptive demand, and linking food aid to population control.[55]

In 1953, Fairfield Osborn authored *The Limits of the Earth*, where he argued that the world is "under the control of the eternal equation—the relationship between our resources and the numbers as well as the needs of our people."[56] This relationship could be expressed in a simple ratio dividing the earth's resources by the number of people on it, and scarcities were provoked by the finite nature of the planet. Beyond their arguments about natural limits, both Osborn and Vogt claimed that resource scarcities and environmental degradation were direct causes of war, and had spurred World Wars I and II. They were concerned about overconsumption of resources and the exporting of American consumerist values in the postwar period, arguing that American standards of living, if spread around the world, would have disastrous effects. Their arguments about consumption and population were inseparable: both were problems that needed to be managed and reduced in order to ensure not only ecological, but also geopolitical, stability.[57] These ideas would be taken up on a national scale a little over a decade later, facilitated by widespread attention to a rapidly growing movement: the mainstream American environmental movement.

Voluntarism, Coercion, and Carrying Capacity as Public Policy

Beginning in the late 1960s, prominent environmentalists turned to a focus on "overpopulation-induced poverty and war combined with new ecological models" to bring about new ways of thinking, talking about, and advocating for environmental issues in the U.S.[58] The independence of formerly colonized nations in the Global South and the Cold War struggle over third-world resources led many Americans to worry about population growth in the Global South as a potential threat to U.S. national security. For others, it was a planetary problem that could bridge regional divides. American scientists Garrett Hardin and Paul Ehrlich were two of the most prominent voices who linked carrying capacity ideas to human population control, arguing that earthly limits, and the ecological crises they produced, made it necessary to foreground population control over development, public health, and voluntary management of one's own fertility. They also publicly returned the population conversation to race and class by including white Americans and the middle class in their notions of shared population responsibility, even as they indirectly invoked the role of race and culture in producing unsustainable population growth in the Global South. While many of their writings supported coercive population control policies, to a limited extent they also promoted advocacy for women-centered reproductive health interventions. In doing so, Hardin and Ehrlich laid the groundwork for current manifestations of sexual stewardship.

In 1968, ecologist Garrett Hardin wrote a well-known article, "The Tragedy of the Commons," in which he described his ideas about the dangers of overpopulation by using the metaphor of a common pasture available to all to graze their herds. His argument was that each person using the commons would graze as many cattle as possible, acting out of self-interest and seeking to achieve maximum possible gains from the land. This, he argued, would ultimately lead to destruction of the land and peril for humans: "Ruin is the destination toward which all men rush, each pursuing his own best interest in a society that believes in the freedom of the commons. Freedom in a commons brings ruin to all."[59] The commons he described symbolized the earth and natural resources; the metaphor of grazing cattle represented population growth. Hardin's aim was to apply the concept of carrying capacity literally, and he concluded that there is no technical solution available; instead, the population problem was one that required moral solutions. This conclusion sprang "directly from biological facts"[60]; he was concerned with identifying the optimum population size, the central problem-question of carrying capacity.

The article was a "philosophical defense of coercion so influential, especially in environmental circles, that it was called the 'Magna Carta' of compulsory population control."[61] The tragedy of the commons as he conceived it is actually the tragedy of the welfare state: Hardin argued that if families only depended on their own resources and the children of "improvident" parents starved to death, there would be "no public interest in controlling the breeding of families."[62] His solution to the problem? A sense of moral responsibility through "mutual coercion, mutually agreed upon" by the majority of affected people, but particularly focused on women.[63] Hardin was concerned that technical solutions to the population problem would not work, because they did not impact people's values. The freedom to reproduce, or breed, as he termed it, would bring ruin to all—as such, coercive population control was a necessary strategy to save everyone.

Moreover, Hardin's ideas actually arose from virulent anti-immigrant sentiment. These politics came to the forefront in his 1974 essay "Living on a Lifeboat," in which he argued that the problems of the tragedy of the commons were no longer simply urgent, they were now a matter of human survival. Likening the regions of the world to lifeboats on a hostile and dangerous sea, Hardin described the rich populations of the Global North living in lifeboats surrounded by dangerous waters. The poor, living on their own, far more crowded boats, were constantly falling overboard and hoping to be taken in and saved by the people on the rich boats, which were limited in capacity. Hardin argued that admitting impoverished "others" to the wealthy boats, knowing the boats' limits, would capsize the boat and kill everyone on board—an unethical and unacceptable solution. As a remedy, he proposed the ethics of the lifeboat, to "admit no more to the boat and preserve the small safety factor. Survival of the people in the lifeboat is then possible (though we shall have to be on our guard against boarding parties)."[64]

This neo-Darwinian argument formed the basis of several population control policies Hardin proposed, which included cutting food aid to less developed countries and severely restricting immigration to the U.S. He also brought women under the banner of his population control agenda in contradictory ways, alternately arguing a need to control women's childbearing, and for an expansion of their access to reproductive services. Hardin wrote, "The Women's Liberation movement may not like it, but control must be exerted through females" because divorce and remarriage constrained the responsibilities of couples and men.[65] At the same time, Hardin was the first to invoke the term "abortion on demand," a phrase which would later become a rallying cry for reproductive rights groups.[66]

Hardin's contemporary, biologist Paul Ehrlich, argued similarly that starving populations in the Global South were beyond the scope of concern of wealthier nations, doomed as they were to an inevitable destiny of starvation based on rampant population growth. His well-known text, *The Population Bomb*, opens thus:

> The battle to feed all of humanity is over. In the 1970's the world will undergo famines—hundreds of millions of people are going to starve to death in spite of any crash programs embarked upon now. At this late date nothing can prevent a substantial increase in the world death rate, although many lives could be saved through dramatic programs to 'stretch' the carrying capacity of the earth by increasing food production. But these programs will only provide a stay of execution unless they are accompanied by determined and successful efforts at population control. Population control is the conscious regulation of the numbers of human beings to meet the needs, not just of individual families, but of society as a whole.[67]

Ehrlich strongly advocated for the U.S. and other industrialized countries to cut food aid to poor countries that were deemed "beyond help," unless they adopted national population policies predicated on universal use of contraceptives. He also advocated population control in the U.S., "hopefully through changes in our value system, but by compulsion if voluntary methods fail,"[68] in addition to exhorting Americans to change their lifestyles and consumption practices to lessen their impact on the world's resources. Using both an intellectual lens and an emotional one, he appealed specifically to a white, middle-class American audience, stoking their fears of the growing global presence of dark-skinned others. This emotional lens was refracted through his description of a family trip to India, on a "stinking hot night in Delhi," which he famously described through his visceral reaction to the surroundings. As their ancient, "flea infested" taxi crawled through the streets, the Ehrlichs entered a slum area where the streets were "alive with people. People eating, people washing, people sleeping. People visiting, arguing, and screaming. People thrusting their hands through the taxi window, begging. People defecating and urinating. People clinging to buses. People herding animals. People, people, people, people."[69] The scene felt to him like hell; he and his family were frightened of the endless swell of bodies all around them. This was a sign of the hell that could afflict Americans if overpopulation were allowed to proceed unchecked.

Unlike others before him, Ehrlich also argued that the U.S. and other "overdeveloped" countries are also overpopulated because they cannot produce adequate resources to maintain their populations' affluent lifestyles of resource consumption and technology use. All nations of the earth were overpopulated, according to Ehrlich, and in need of population control; the kind of overpopulation was divided along the lines of resource shortage and population boom versus overconsumption of resources by small and affluent populations.

Further, Ehrlich actually challenged the family planning solutions proposed by development actors and demographers, arguing that these approaches failed to fully resolve the environmental problems created by population growth. He broke with demographers over the issue of modernization and technology: modernizers looked for technological solutions to what they saw as cultural problems, whereas Ehrlich saw those problems as biological and thus not amenable to technological changes.

Ehrlich was a complicated figure—just as he emphasized the "frightening" growth of poor, racialized populations overseas, he also turned the lens back on the white middle and upper classes in the U.S. As a strong antinatalist, he critiqued the reproductive and consumption practices of whites, as well as critiquing racial injustice and race-based justifications for population control in the U.S. Just three years after publishing *The Population Bomb*, Ehrlich and his co-author Richard Harriman wrote a book titled *How to be a Survivor: A Plan to Save Spaceship Earth,* in which they laid out their arguments in terms of race and civil rights. First and foremost, they proclaimed, white middle class and wealthy Americans should be the primary targets of population control efforts, and the government should offer maximum incentives and minimal coercion unless the incentives proved unsuccessful, in which case more coercive measures should be undertaken. More to the point, these interventions were not meant to target marginalized communities: "Above all, no effort should be made to single out the poor, people on welfare, or nonwhites as special targets for population control . . . it is among the middle class and the wealthy population that population growth presents the most serious problems."[70]

Why did Ehrlich and Harriman take this stance? For them, despite the fact that population growth rates were higher among nonwhites and the poor, it was affluence-based resource consumption in the U.S. that was destroying the environment. Further, they argued that affluence-based consumption is directly connected to the environmental burdens of the poor: higher body burdens of DDT and other pesticides are attributable to overuse and misuse of these chemicals by commercial agriculture and other industries, as well as the lower quality, nutrient-deficient foods available in low-income communities.

Ehrlich and Harriman also thought of focusing on middle and upper class whites as a strategic way of minimizing racial tensions and controversies, arguing that "the best way to avoid any hint of genocide is to control the population of the dominant group."[71] They were concerned about black communities' anxieties about coercion in contraceptive and abortion clinics in "ghettos"; thus, they proposed that reproductive health services be administered locally and voluntarily by community members: "Black women in a ghetto should have access to contraception and safe abortion *if they want it*. But they should not have to seek such services from white men or women. Let black women run the clinics."[72] Further, their propositions on race included increasing living standards for the poor as a way of addressing white racism. For example, they argued that contraceptives and abortion should be made freely available to all American women, and that those whites who were uncomfortable with the thought of a rising black population should do their best to support increasing affluence among blacks, since affluent blacks tend to have fewer children than affluent whites. Of course, Ehrlich's positions on population policy were complex and contradictory: he advocated coercive state-based proposals (taxes on child-related necessities, sterilizing all males with three children or more in high-growth nations, adding sterilizing chemicals to the U.S. public water supply) that would have had disproportionately impacted people of color and the poor.[73]

While producing their social agenda on population, Ehrlich and his contemporaries were also steeped in the production of scientific models and mathematical equations, attempting to ground their dire environmental predictions and socio-political advocacy in cold, hard facts. In 1971, Ehrlich and John Holdren developed an equation, I=PAT, which they claimed demonstrated the ways population interacts with affluence, or resource consumption, and technology use to produce a range of negative environmental impacts. In this equation, I (Impacts) are the direct result of the product of P (Population), A (Affluence), and T (Technology), a model that reduces a host of uneven and complex social, political, and economic factors determining population trends and resource use into a rather extreme level of simplicity. These elements did not represent equal relationships to environmental impacts; rather, the authors argued that population growth "causes a *disproportionate* negative impact on the environment,"[74] a factor that they promoted as a clear, scientific basis for advocating population control. Moreover, the I=PAT model was intended to represent a universal set of relationships, with a one-size-fits-all approach focused on population control, changing systems of technology distribution, restricted resource use, and poverty alleviation at a global level.

The twentieth-century proliferation of carrying capacity and planetary limits discourses, as well as the quantification of political arguments about population growth, culminated in the Club of Rome–commissioned project, *The Limits to Growth*. This book drew on system dynamics theory to produce a series of computer-generated models projecting how the exponential growth of population, food production, and consumption patterns would interact with resources such as petroleum, gold, iron, and chromium over a period of two hundred years. The computer models rather predictably indicated that eventually population growth would overshoot available resources and a collapse in food production would occur, leading to the Club's conclusion that human population growth and resource use far exceed the carrying capacity of the earth's finite resources. The book's arguments were focused on resource shortages and scarcity—their conclusions were that governments would divert resources to managing scarcity such that overall quality of life for many people would decline in the twenty-first century. Like Hardin and Ehrlich, theirs was a clear carrying capacity argument; however, it was buttressed by the technologically advanced strategy of extending predictions into the near and distant future through scenario modeling (see chapter 3). In an updated version published thirty years later, the authors repeated their dire prediction: "Sadly, we believe the world will experience overshoot and collapse in global resource use and emissions much the same as the dot.com bubble—though on a much longer time scale."[75]

Their proposed solution to overshoot and collapse was sustainability, defined as the conditions that allow societies to persist over generations. Unlike neo-Malthusians before them, they argued for a modicum of social equality; their notion of sustainability required keeping exponential growth in check through providing universal, adequate standards of living and fair distribution of resources. They advocated limited growth focused on supporting important social goals and enhancing sustainability, and exhorted societies to eventually stop pursuing this growth once their specific goals were met. This was not necessarily zero growth, but rather a strategy based on qualitative assessments of who and what development is for, who would benefit, what the costs would be, how long it would last, and whether the earth could accommodate it. In this model, sustainable societies are also societies that address and overcome poverty, unemployment, and unmet nonmaterial needs.

Conclusion

As this chapter has demonstrated, population has been constructed over time as a scientific, social, political, and economic problem. Throughout the twentieth century, ideas of national identity, geopolitical stability, natural resource security, and living with limits have preoccupied various scientists and policymakers, directly influencing their ideas about how these scientific ideas should be applied. The next chapter turns to these questions in more depth. It explores the close linking of science and politics in knowledge production on population growth and climate change, and demonstrates how this relationship entangles and embeds questions of power and privilege operating from behind the scenes. Moreover, it explores and raises questions about the role of policy concerns in scientific research that not only projects, but produces, the future.

3

Scientists, Donors, and the Politics of Anticipating the Future

Most of us are interested in knowing the future. We want to know the future in order to control it. We want to control the future in order to benefit from it, or to mitigate or avoid harms that we would otherwise suffer.
—Dale Jamieson, 1998

Anticipating the future has always been central to population-environment knowledge, activism, and policymaking. As Adams, Murphy, and Clarke note, "Anticipation is not just betting on the future; it is a moral economy in which the future sets the conditions of possibility for action in the present, in which the future is inhabited in the present."[1] Anticipation is a form of affect; it links the tangible present to the possible future by paving the way for actions that can bring any range of futures into being. For population advocates, the linked futures of climate change and population growth are very much in the present, orienting scientific knowledge production, crisis discourses, and advocacy strategies with long-term implications.

Whether advocating population control or SRHR, environmentalists in the U.S. have a long tradition of drawing on scientific data to ground their arguments. Chapter 2 demonstrated the ways that these scientific ideas are deeply entwined with political, economic, and cultural arguments; in this chapter, I argue that, while the public face of populationism has been dominated by scientists and activists, donors have long played a key advocacy role from behind the scenes. In fact, we cannot actually understand the paradigms and priorities of scientists without understanding the institutional funding mandates, political priorities, and personal politics that shape scientific research paradigms. Exploring the role of donors and their close, long-standing relationships with scientists offers a means of understanding how and why science operates as a tool of advocacy. A central argument of this chapter is that public and private donors in the U.S. act as agents of change with scientific, social, and political agendas of their own.[2] As this chapter will demonstrate, individuals within donor institutions act as population-climate policy advocates via grant funding, in which they create the conditions for researchers and activist groups to shape public debates. They also play a powerful role in how key actors anticipate, envision, and encourage others to act on possible climate and population futures.

Over the course of conducting the research for this book, I discovered that for some scientists, particularly those working at the intersection of population growth and global environmental change, their projection models originate from twinned desires: the desire to produce more knowledge about planetary conditions, and the desire to intervene on them. These desires cannot be teased apart; they infuse everything from research funding, to the kinds of models that are produced, and the policy questions they lend themselves to. In this chapter, I argue that the combination of knowledge production and advocacy constitute an *anticipatory politics*, in which multiple possible futures are projected for the purpose of opening possibilities for intervention—intervention that can create some of those futures and prevent others through policy change.

Population Science, Climate Change, and Development in Africa

Around the turn of the millennium, scientists began to circulate a small body of research at the nexus of greenhouse gas emissions (GGEs) and human numbers. Multiple projects have claimed that slowing future population growth would make a significant impact on GGEs. A 2001 book proposed that "in LDCs, policies such as voluntary family planning programs and investments in girls' education are not only desirable in their own right, but also accelerate fertility decline, which may have significant benefits in the context of climate change."[3] In an earth systems study, scientists argued that "by the end of the century, the effect of slower population growth would be . . . significant, reducing total emissions from fossil fuel use by 37–41%."[4] The study concluded that slowing global population growth overall can serve as a key climate change mitigation strategy, reducing carbon emissions by up to one million megatons by 2100.

Other studies published around the same time offered somewhat different perspectives. For example, biostatisticians Murtaugh and Schlax analyzed the "carbon legacies" or projected lifetime emissions of individual women and all of their future progeny, comparing them by country and region. They found an inverse relationship between individual childbearing and per capita GGEs; in other words, countries where women bear the fewest children are most often those with the highest rates of per capita GGEs and the highest carbon legacies. According to this model, the average American woman's carbon legacy is more than 85 times that of an average woman in Nigeria, a country with a much faster population growth rate than the U.S.[5] In another study focused more closely on energy use, an urban development researcher found that over a fifty-five-year period, nations with rapid population growth had little GGE growth. Rather, his analysis demonstrated that GGEs were driven by the growth in *consumers* and levels of consumption across world regions.[6] A more recent study by economists confirmed these results, finding that changes in gross domestic product (GDP) were the closest proximate drivers of carbon emissions. The study found no relationship between short-term world population growth and emissions.[7]

A third perspective has also emerged, one based on using these future-oriented data to argue for better approaches to development today. Given that the problem of GGEs is primarily one of resource consumption, not human numbers alone, and that demographic changes cannot be reversed overnight through contraceptive distribution, researchers argue that what is needed is a different approach to development—one in which urban growth is designed around energy efficiency, reducing emissions, and providing adequate housing for the poor.[8]

Some Global South leaders share this approach. In 2001, the United Nations Framework Convention on Climate Change (UNFCCC) Conference of Parties established the preparation of National Adaptation Programmes of Action (NAPAs) as a strategy for the forty-nine least developed countries (LDCs) to develop plans addressing the adverse effects of climate change, and to promote adaptation at a national scale. In a 2009 study conducted by Population Action International[9] (PAI), forty-one NAPAS were analyzed for their population and reproductive health content. While the majority of the plans identified rapid population growth as exacerbating vulnerability to climate change, only six identified slowing population growth as a priority adaptation action, instead favoring development approaches addressing food insecurity, access to clean water, basic health needs, and education. Nevertheless, PAI concluded its analysis by arguing that "NAPAs should translate the recognition of population pressure as a factor related to the ability of countries to adapt to climate change into relevant project activities."[10]

Why would PAI ignore the conclusions outlined by LDC leaders in the NAPAs? One clear answer is that the conclusions did not align with PAI's projected vision of the future. At the 2010 COP 16 meetings in Cancun, PAI launched a new digital project, *Mapping Population and Climate Change Hotspots*.[11] The project produced a set of interactive global maps depicting population dynamics, including growth rates and unmet need for family planning, which could then be layered with elements such as "projected changes in agricultural production," "water scarce or water stressed countries," or "resilience to climate change" to produce a visual representation of climate "hotspots." The model defined hotspots as countries or regions with high rates of population growth, high projected declines in agricultural production, and low resilience to climate change. The maps offered a visual narrative of the link between family planning and climate change, with stark, colorful contrasts depicting hotspots both now and in the future. Moreover, all of the countries depicted as hotspots were in sub-Saharan Africa. Was Africa the only region facing high population growth and climate change vulnerability? Why was the narrative in the maps built on what appeared to be African exceptionalism?

One answer emerged in a special PAI report from 2012, *Population Dynamics, Climate Change, and Sustainable Development in Africa*, which opens with the following:

> SSA's population is growing more rapidly than other regions of the world. Rapid population growth and climate change are speeding up the region's environmental degradation. This makes people more vulnerable to climate change impacts and undermines sustainable development on the continent. Development efforts in several countries in SSA are harmed by a combination of high rates of population growth, high projected declines in agricultural production and low resilience to climate change.

> We classify such countries as population and climate hotspots. In these hotspots, addressing population challenges will help increase resilience to climate change, and contribute to development goals such as better food and water security.[12]

The report depicts a continent in which high population growth and increasing urbanization render populations vulnerable to the intersections of poverty and exposure to the increasing effects of climate change. Sea level rise, flooding, increasing storm activity, food insecurity due to drought, reduced crop yields, and climate-induced water scarcity were all framed as heightening the continent's vulnerability in a region already acknowledged to be the least resilient to climate change. The proffered solution was, of course, family planning: "Given the strong links between population and climate change, tackling the issues jointly at the policy and program levels makes sense. Looking at rapid population growth and climate change risks together would help identify groups of people who are vulnerable to these twin challenges and illuminate how to help them adapt."[13]

Designating African countries as hotspots in need of immediate population-climate interventions is reminiscent of what Hartmann refers to as the Malthusian Anticipatory Regime for Africa, or MARA.[14] She argues that discourses constructing Africa's singularity as an impending population and climate disaster make a case for population interventions, while American and European defense interests identify the region as a site of concern for future climate-driven conflict. MARA then not only legitimates but necessitates a range of militarized interventions to respond to the urgency of the continent's population and climate crises, while securing geopolitical stability for the region and the world.

MARA and similar kinds of discourses play on old racialized fears of the impoverished, dark-skinned, rural poor migrating to the industrialized north, driven by environmental scarcity and conflict. Climate conflict and climate refugee discourses draw heavily from the work of Canadian political scientist Thomas Homer-Dixon, whose work joined neo-Malthusian arguments with systems theory to produce theories of violent conflict, political instability, and large-scale population migration resulting from resource scarcities. These arguments have historically been just as influential in informing neo-Malthusian family planning advocacy as women-centered arguments have; in fact, despite the strong emphasis on women's reproductive rights and empowerment in the Cairo Consensus, environmental security debates played a prominent role in guiding some of the discussions held at the ICPD.[15]

Climate scarcity and conflict framings are effective in mobilizing development narratives because they are rendered through the lens of possible, and undesirable, futures. As the introduction chapter discussed, individual childbearing has been declining across the continent, though unevenly, for decades. The rate of decline is slower than other regions of the world, leaving Africa an outlier in terms of population growth rates. However, annual GGEs on the continent are among the lowest on the planet. Given this context, it is strange that Africa would be prioritized for population-climate interventions—unless one is focused on the future. In projection models, the continent's future signals the dystopian narratives of war, unrest, and mass migration. However, these projections not only *model* the future: given their power to mobilize human and financial resources for contraceptive and military interventions, they are *making* futures.

Modeling and Making the Future

Future-based projection models are foundational to how we understand both population and climate change. Probabilistic projections are just that: projections of probabilities. They depict in visual form where global population or GGEs *could be* if certain conditions are met—conditions like specific fertility rates or life expectancies at birth, or use of high-carbon energy sources. They are visual representations of "if-then" statements; if certain assumed conditions are met, then these are likely outcomes. As a result, the projections are usually depicted in a range of possible outcomes, rather than settling on one solid figure.

Beginning in 1990, the Intergovernmental Panel on Climate Change (IPCC)—the scientific research arm of the leading international climate change treaty organization, the UNFCCC—has produced a series of reports covering technical, social, and socio-economic information underpinning human-driven climate change. After collecting and analyzing data from thousands of published climate-related studies around the world, the IPCC prepares Assessment Reports reflecting the current state of knowledge on climate change. The data are used to produce projections of GGEs into the future; as with population projections, the year 2100 is a commonly modeled target date. In fact, since the year 2000, the IPCC projections have not only been similar to population projections—they have *included population projections* in their scenarios.

In 2000, the IPCC produced its Third Assessment Report (TAR) on global climate change. It was the first time that the TAR included a new Special Report on Emissions Scenarios—scenarios being defined as "projections of a potential future"—in which four narrative storylines depicting "different demographic, social, economic, technological, and environmental developments that diverge in increasingly irreversible ways" could be explored.[16] The first two storylines, A1 and A2, depicted either a world of rapid developments in economic growth and technology, and peak population in 2050, or a world of continuously growing global population and slow, inconsistent economic growth. The B1 and B2 storylines mapped onto the A storylines in terms of population, but with different economic structures. B1 depicted the same population trends as in A1, but with a focus on a service and information economy and clean technology; B2 reflected continuously increasing population, and a global focus on local environmental, economic, and social sustainability solutions.[17] In the projection models, the growth curves of population and emissions mapped neatly onto each other; the visual narrative seemed to say that population and emissions grow in tandem.

Are these models a prediction of the future? Do they offer a blueprint or road map to the potential solutions that can prevent social, economic, and ecological disaster? In a sense, it would appear that they could. Models such as these provide a particular way of understanding the future by visualizing it. Once it is visualized, mapped onto the smooth lines of a computer graph model, it appears intervenable, something that can be controlled. The host of complex factors, elements, data, not to mention vast uncertainty, that go into producing these images are smoothed into clean lines and curvilinear graphs. However, projections are not predictions. A prediction is a "statement that something *will happen in the future* based on what is known today," whereas a projection is a "statement that *it is possible that something will happen in the future* if certain conditions develop."[18] This chapter is concerned with the latter. The focus on population growth and climate change is based on a focus on *possible* futures, brought into being through the production of scientific scenarios. However, these futures are not simply the vision of the scientists who model them: they are the result of complex relations between scientists, the donors who fund them, and the advocates who drive policymaking.

Climate change mobilizes a constellation of future-oriented practices: political activism, scenario modeling, and exhortations to change energy-using behaviors, among others. Current actions take on a sense of urgency, suggesting multiple possible futures ranging from the optimistic to the profoundly dystopian. This is not a new way of understanding environmental problems or population trends: as chapter 2 demonstrated, many American scientists studying population and the environment throughout the twentieth century were motivated by concerns with resource consumption, depletion, and scarcity. They were focused on dystopian environmental futures, which framed their understandings of global population growth, as well as resource use. While the hyperbolic language found in population-environment writings of decades past has been muted, scientists working at this nexus today are still somewhat driven by a focus on dystopian possibilities—and these narratives drive the production of policy-oriented science, a necessary resource for mobilizing alternative outcomes.

Anticipatory Politics

In April 2010, an Earth Systems scientist gave a research presentation and public webinar on population and climate change at the Woodrow Wilson Center in Washington, DC. The convening was organized to brief members of the international development policy community on the results of an innovative multi-year research project in the arena of demographic projections and GGEs. Sitting at the nexus of climate science and demography, the project was somewhat unique in engaging tools and methods from both disciplines, as well as offering recommendations for policy actors outside of the climate arena—specifically, those working in the field of population and family planning.

The scientist began by describing demographic factors like population growth and size, age and urbanization, as well as less traditional demographic categories such as educational attainment and population health, as key determinants of GGEs and human impacts of climate change. His work was based on a projection model based on the IPCC Special Report on Emissions Scenarios, integrated with recent population projections from the United Nations, and disaggregated by world regions and major emitting countries. When integrated, the model formed a smooth visual image of possible population and emission futures, ranging from the lowest to the highest possible projected human numbers and emissions.

The model's visual appeal belied the interpretive challenges it offered to non-specialists; however, the scientist cut quickly to the chase, stating that over the next hundred years or so, human population growth in countries of the Global South would become a significant source of GGEs, thus requiring a prioritization of population interventions in the present:

> If I could take a leap and summarize the conclusions here in a sentence, it would be that if you did slow population growth, it would likely reduce greenhouse gas emissions significantly in the long term. Also, slower population growth would ease adaptation to climate change as well. Therefore from a policy perspective . . . policies that have the effect of leading to lower fertility and to slower population growth can be considered win-win from the climate point of view. The first win is that there are plenty of good reasons for those policies that have nothing to do with climate or environment that have multiple benefits in their own right. But because through slowing population growth, they would likely make the climate problem easier to solve, there's a double benefit. This hasn't caught on and gained traction in either the policy or research worlds.

His study argued that slowing population growth overall in the long run would reduce emissions and ease the path to adaptation, while growing urbanization throughout the Global South mitigated against this reduction. In summing up his presentation, the scientist interjected a bit of nuance: he was careful to note that demography is not the sole nor the most important element that impacts emissions; rather, technology is. He argued that slower population growth could contribute significantly to lowering emissions, but would not solve the problem, "nor is it a main factor. You can't force fit that into a model."

The audience, however, was less interested in technological change than in new ways of communicating populationist urgency. During the Q&A following the presentation, it was clear that anticipatory politics were at the forefront of the development crowd's reflections on the science:

> Q: The issue hasn't gained much traction. You've been at this for a while; are we gaining traction? And at what levels? If not, what needs to be done?
>
> A: I have less experience with this on the policy side. I'm guessing that if there were a more solid and sophisticated and comprehensive basis on the science side for drawing conclusions about what the possible effects might be, that might facilitate the policy process, but I don't know. There might be other priorities. Population is a sensitive issue and I don't think one study or even ten is going to change that. I hope that there will eventually be a better conservation about this. I do think that's changing. Before you couldn't even raise the issue because it was so sensitive. So I'm hoping that this kind of work can inform things and say probably as best we can tell, it would help solve the problem. However, just because something can help solve the problem doesn't make it a good idea. The policy debate needs to take a lot of things into account, including value systems.
>
> Q: Do you feel comfortable with population advocates using your results to make a case for increased funding for family planning?
>
> A: Yes. But I'm also comfortable with others making the alternative case and saying that it's . . . not enough.

While this particular scientist would not draw heavy-handed conclusions from projection models, some population-climate studies have been more direct about communicating a need to take immediate action. In late 2009, a British charity and think tank known as the Optimum Population Trust (OPT) launched a now-defunct project named PopOffsets, proclaiming itself the world's first project that provided individuals and organizations the means to offset their carbon footprint via financial support to family planning projects. According to their calculations, contributing seven U.S. dollars through their website would provide family planning services to women; through providing the contraceptives, fewer babies would be born, and less consumers would use carbon-emitting resources. As the project website claimed, addressing the unmet need for contraceptives was "the lowest cost way of reducing CO_2 emissions and climate change . . . without any environmental downsides."[19]

PopOffsets's project was based on the concept of carbon offsets, a strategy to manage greenhouse gases by compensating for emissions in one area through reductions in another. The basic logic behind it is that investing in projects that reduce greenhouse gases elsewhere, in this case through family planning and the prevention of new births, would be "cheaper, easier, and faster than domestic reductions, providing greater benefits to the atmosphere as well as to sustainable development, especially when offsets involve projects in the developing world."[20] These assertions arose from an OPT report released several months earlier, in which the author conducted a cost-benefit analysis linking population reductions and CO_2 emissions, and assessed the cost effectiveness of universal contraceptive access on carbon emissions reductions between 2010 and 2050. The rather simplistic conclusion was that "fewer people will emit fewer tonnes of carbon dioxide."[21] The author of the report reached this conclusion through basic math: he estimated the cost of providing contraceptive access to all women who had an unmet need for contraceptives, and analyzed the results against projection models of population trends and carbon emissions. From these models, the author concluded that for each $7 spent on family planning provision, you could reduce future greenhouse gas emissions by more than one ton (assuming all unmet need was met between 2010 and 2050). In contrast, the cost of a one-ton reduction achieved through the use of low-carbon technologies was estimated at a minimum of $32, a $25 increase over the cost of similar reductions achieved through family planning.

This was based on a particular calculus assigning values to human lives, contraceptives, and tons of carbon, and asserting a symbolic equivalency between them. Carbon offsets were developed as part of the international climate treaty of 1997, the Kyoto Protocol, which established a mechanism for industrialized nations to invest in clean energy projects, such as reforestation and biofuels plantation projects, as a means of trading carbon credits. These mechanisms are predicated upon a particular commodification of nature—one in which isolated and abstracted elements of nature become valued and tradable in relation to other isolated and abstracted elements, as a proxy for carbon emissions. In the PopOffsets project, human lives became part of the calculus, not through existing life, but through an assessment of the carbon-emitting value of lives that can be averted. The value of averted lives is calculated, assessed, and offered to Northern consumers as a means of offsetting their polluting behaviors. This logic transforms humans into potential humans, potentially averted humans, and ultimately potentially averted emissions.

The logic of averted-humans-as-averted-emissions raises a host of questions about the power dynamics of how value in human life is assigned, to whom, and at what scale. Historically, science has been complicit in schemes to assign value to life in ways that "calibrate and then exploit the differential worth of human life" for economic aims.[22] To borrow a phrase from Murphy, this reflects strategies of the *economization of life*, in which some lives are deemed investable and others are expendable and avertable.

This discourse of averted-lives-as-averted-emissions appeared on the world stage at the 2009 COP15 in Copenhagen when a Chinese delegate, the Vice Minister of the National Population and Family Planning Commission, stated that "dealing with climate change is not simply an issue of CO2 emission reduction but a comprehensive challenge involving political, economic, social, cultural and ecological issues, and the population concern fits right into the picture."[23] Referring to China's one-child policy,[24] the Minister argued that 400 million births had been averted in China, which she estimated as having saved 1.8 billion tons of CO2 each year. Claiming a moral authority on the basis of national policies leading to averted emissions, the Minister argued that China's population policy had provided benefits across sectors, and at a global scale. In this case, a state which has long been critiqued for its coercive national population policy was able to use the calculus in human lives to position itself as an environmentally responsible global actor. The appeal to the averted-humans-as-averted-emissions calculus facilitated an avoidance of the moral and ethical questions and human rights abuses under the one-child rule. Instead, even coercive population control could emerge as an example of ethical environmental practice.

As the China example demonstrates, the calculus in human lives potentially makes ethically untenable practices possible by reducing human lives into a set of calculable values and rendering them avertable in the service of broader environmental goals. The power dynamics of calculation—who is able to calculate, who is calculated, and ultimately whose lives are deemed avertable—are at the heart of population interventions and advocacy.[25] Again, these calculations rest on quantified projections of the future, projections that arise both from scientific inquiry and from the donor institutions that make their work possible. The remainder of the chapter explores these dynamics through the in-depth treatment of one case in which donor advocacy and scientific innovation combine to produce new possible population futures.

Steering the Ship: Donor Advocacy

In May 1998, the manager of a population-environment program at a Washington NGO sent a ten-page letter to Pamela,[26] a former donor at a small, private philanthropic foundation in the Silicon Valley who funded his program. His letter outlined what he described as a crisis of legitimacy in the field of population-environment advocacy and intervention, and he proposed that she and other population donors rectify the problem through funding more science. The letter read, "The greatest need may be to identify ways to fund careful, accurate research, worthy of the widest dissemination, which is at least not hostile to the linkage between population dynamics and environmental problems." He claimed this would aid in countering the existing "contrarian bias" on the part of critical academic scholars, which seemed to be directed toward the purposes of "shatter[ing] 'shibboleths' about population and the environment [rather] than . . . clarify[ing] the precise nature of the linkage."

As the writer saw it, advancing new scientific knowledge would facilitate "developing new ways to reach the public with an enhanced and respectable presentation of the population-environment linkage," which would then bolster advocacy efforts by attracting environmental journalists and other writers to take an interest in population issues. This would in turn help to confer a sense of scientific legitimacy on population-environment activism. He called upon Pamela and other donors to support academic research that was "at least open to the hypothesis that population dynamics are decisive in the expansion and worsening of environmental problems," noting that a number of scientific academies had previously issued similar statements. Of note, he observed that these statements were not based on consensus approaches, nor on actual research findings, but rather "on the commonsense understanding of scientists of various disciplines."

The conundrum he laid out was clear: population-environment advocates in environmental organizations needed more solid scientific ground from which to make their case for population advocacy. Rather than waiting for scientists to produce this research on their own, he identified foundations, with their autonomous access to private funds and freedom from public scrutiny, as the most appropriate solution to the problem. At the same time, the heart of the request was clear: in order to produce the kinds of scientific results he was soliciting, Pamela would need to engage in donor advocacy by supporting value-driven science.[27] To address this challenge, the writer suggested a simple solution: the "selection of institutional homes and individual researchers known to be of the highest academic and intellectual caliber combined with the courage to challenge conventional scholarly views," thus providing a stronger evidence base linking population growth and environmental degradation, while grounding one's arguments in academic and institutional legitimacy.

Pamela was listening. A long-time proponent of global population stabilization, particularly in Africa and Asia, Pamela was trained in ecology, and had long been concerned about strengthening the scientific argument linking population growth to environmental degradation in the Global South. The letter's request also dovetailed with an ongoing quest in the population-environment donor community to strengthen the scientific and political bases for population-environment advocacy, particularly in the context of declining public funding for integrated approaches.[28]

Several years later, Pamela had the opportunity to fulfill the request. She contacted a senior demographer at the Population Council, one of the largest population research organizations in the U.S., asking him to recommend a scientist who could conduct the kind of research that would not only make a strong scientific case linking Global South population growth to GGEs and climate change, but that would also provide a justifying framework for advocacy work. The demographer had just the person: a modeler with a long-term research agenda focused on modeling the impacts of population growth on future GGEs. In a subsequent e-mail to the scientist, Pamela described her foundation's commitment to population and environment issues, as well as outlining the potential interest in funding population-climate change research:

> The Foundation . . . has a long-standing interest in both environment and population issues, including the interrelationships that exist between these two fields. Climate change is one of our current environmental areas of interest, and increasing resources for international family planning and reproductive health services is a major focus of our population program . . . I am writing to inquire about whether [your forthcoming report] will include any mention of population growth, or any recommendations for increasing funding for reproductive health services in the LDC's [Least Developed Countries] as part of a strategy for adapting to climate change. If so, we might be quite interested in supporting efforts aimed at the broadest possible public dissemination and discussion of that portion of the report.

Her offer message was well received; the scientist responded the next day with a detailed list of several nascent projects that he was pursuing, work that could immediately be expanded with her foundation's support. With that, a new donor-grantee relationship was born.

These interactions shed light on the everyday, behind-the-scenes processes in which powerful actors co-produce scientific knowledge and policy-oriented practice. Pamela had been searching for a way to bring a scientific grounding to a long-standing interest of hers: a possible set of relationships between population growth in low-emitting countries in the Global South and future climate change. Without projection-based research, the relationship between high population growth and GGEs could not be established. But for Pamela, personal interest, established institutional priorities, and access to private money entwined to drive population-climate linkages forward. In an interview, she described her role as facilitating scientific innovation:

> We were a very, very early funder in this area. We identified it [the impact of population growth on greenhouse gas emissions] as a potential inter-

> est 2.5–3 years ago and we looked for who would be the right scientists to fund to demonstrate whether or not it did matter whether the population grew in terms of impacts on climate change. The accepted perspective at the time was that 95% of growth would happen in the poorest countries of the world, and that those folks didn't have cars and their carbon emissions were very small. We weren't sure that was true; we wanted the science to know more.

I visited Pamela in her office in early 2010. The bookcases lining the walls were filled with books on conservation, global environmental politics, women's sexual and reproductive health, climate change, and international development. Her file folders spilled over with scientific articles and research briefings, project reports, funding strategy documents, and historical material on the foundation's early population projects focused on reducing immigration from Mexico. Talking with Pamela was a study in contrasts. Her mild-mannered demeanor belied what others described as a relentless, even zealous, focus on population and global environmental catastrophe. Yet, as she and I spoke, it became clear that Pamela situated population-environment linkages in the contexts of both scientific fact and strategic opportunity.

When Pamela described her effort to engage the scientist, it was clear that she did not see herself as engaging in personal or policy advocacy, *per se*. Rather, she saw herself as engaging in scientific advocacy—utilizing her access to financial resources to bring forward innovative research that would have otherwise not been produced. She also served as a conduit for the scientist to access other sources of private funding when he wanted to expand his research:

> We made a grant to him to determine whether or not you could determine whether it made a difference to greenhouse gases if the population was 8 or 10 or 12 billion. He reported that it did make a difference, and that the difference was measurable. But he felt that the report would take longer and require more money to really produce definitive conclusions. So we introduced him to another, much larger and better funded foundation, and they funded him. The research will be out soon; it's compelling and interesting. This puts us way ahead of the curve of foundations.

Private foundations are uniquely positioned to fund research in ways public agencies cannot. A former donor from the Office of Population and Reproductive Health (OPRH) at the United States Agency for International Development (USAID), who considered himself to be a strong proponent of population-climate interventions on scientific grounds, expressed deep skepticism over whether population-climate interventions will ever receive public funding, due to a general lack of scientific knowledge in the American public. He argued that the general public is unaware, scientifically illiterate, and guided by emotion rather than reason, which makes for policy shortcomings: "So I think the weakness of our system is inability to use science to set policy . . . An overarching issue is emotion and politics more so than clear thinking based on our best science."

As a result, he argued, only private foundations can support the scientific knowledge production that many in the American public, to their own peril, refuse to view as important. Foundations have a long history of filling this role in order to facilitate policymaking. Greenhalgh argues that population policies would be untenable without population science: "In policies aimed at governing populations, science-based logics play an especially critical role because population is a biological entity . . . and science claims to be the sole authority on "nature", to which biology belongs. It would be difficult to govern population—or to govern it well—without a science of population."[29] She further argues that "science serves to legitimize both the exercise of power through policy and the authority of the policy makers."[30] The opposite is also true: powerful policy actors exercise a tremendous amount of power in determining what comes to be known as scientific knowledge. We cannot understand the science of population, nor NGO advocacy practices, without the policy-oriented donor context that facilitates their development.

The American philanthropic sector grew dramatically over the course of the twentieth century, from just eighteen foundations before the year 1910, to approximately 50,000 private, corporate, and community foundations by the year 2000.[31] As a result, nonprofits, international NGOs, and community-based organizations have come to be increasingly dependent over the past century on an expanding group of private funding organizations whose scientific, political, and strategic objectives are often closely guarded. In the context of population-environment projects, the process of securing funding from these organizations is somewhat obscure, despite public representations of the grant-making process.

Perusing the websites of the handful of private foundations that operate population programs yields a standard picture of the grant solicitation, application, and funding process. Step 1 involves becoming familiar with the projects that are typically funded by that particular funding agency, including researching grant-making programs, goals and strategies, and geographic specifications. Foundation websites also publish information on recently awarded grants, including the funding amounts that have previously been allocated to grantees across programs so that potential grantees may tailor potential funding requests to fit a set of parameters that will likely yield a successful grant application. This information is also meant to weed out potential grantees that are not likely to be funded, whether due to the amount of the funding request, or because they engage in projects that do not suit the donors' priorities, strategies, or conceptual frameworks.

After a potential grantee has become familiarized with the zones of possibility in grant funding, they initiate step 2: writing a Letter of Inquiry (LOI). At this phase, grant seekers enter into somewhat of a courtship with their potential donors, as they attempt to assess exactly the right phrasing, funding amount, and description of program activities that would achieve the objectives supported by the donor. As one Program Officer I interviewed noted, this process can be a frustrating one for donors in the sense that often grant seekers tailor their LOIs, and later proposals, to what they anticipate to be a successful grant, rather than keeping with their organization's goals and objectives. In step 3, having reviewed LOIs and selected those that appear to best fit the foundation objectives, the foundation solicits full proposals. Only under invitation from the foundation are grant seekers allowed to submit full proposals; those who deviate from this regulation by submitting full proposals outside of these tightly regulated guidelines are disciplined with a refusal of funding. Proposals are made to fit within a set of parameters with respect to length, content, format, and style. Of them, a select number are included in a docket, or a list of funding proposals that are then given final approval by the foundation's board of directors. In this regard, all foundations are different in that smaller organizations have heavy board involvement in funding allocations and docket approval. Larger foundations with multi-million-dollar dockets often have less oversight, and program funding areas are approved as single dockets. After approval of the docket, successful grantees receive a formal letter outlining the terms and stipulations of the award.

I explore the intricacies of the typical process of soliciting and receiving awards from private foundations because a significant portion of private foundation funding for population-environment grant-making has historically deviated from this norm. Population grant-making has followed such a unique trajectory that it can be described as defining a new norm of creative donor financing. As I discovered over the course of numerous interviews, Pamela's actions as an active recruiter of grantees hardly make her an exception to the norms of private donor practice within foundations, particularly in the context of population-environment science and advocacy. Although private foundations are institutions designed to carry out the mandates of their governing boards, individual leadership, personal priorities, and larger political trends may have a significant influence over the direction of particular funding portfolios.

For example, staff members at private foundations are often able to translate their personal interests into advocacy programming through the strategic use of program funds. A donor at a private foundation noted that this personal interest can be central to setting grant-making portfolio agendas: "In the longer term, when [our founder] set up the Foundation, what drove interest in the population field was the environment and the links they saw there. More recently, it was a personal interest of mine, something I was interested in pursuing. It coincided with a former Environment Program Director who has since left. He had approached me and said that there were some clear links between our projects, and that we should think about doing some joint work. It just made sense. So, I came up with our population and climate change work . . ."

Over many years, a number of private foundations, including Hewlett, Packard, Goldman, Compton, Pew, and Summit have approached environmental organizations and offered them sums of money, ranging from the tens of thousands to the hundreds of thousands, to develop population projects.[32] Many donors see their funding efforts as innovative development work, particularly in the context of cross-cutting or multi-sectoral work, given the fact that most donor funding is siloed into single-issue grants. A former donor at a small Washington, DC, foundation told me that private money allows some donors the flexibility to create new synergies between different program streams:

> I ran an experimental initiative to do cross cutting grant-making. I created that program, which centered on creating grassroots programs linking women's reproductive health with conservation and the environment. I also worked on grants that went toward organizational work. My SOW [scope of work] was to do experimental grant-making. But my background was on the role of women in environmental development. The main population strategy at the time was family planning service delivery; the conservation strategy was biodiversity. My question was can we pursue both goals in a synergistic way that meets the needs of both. I wanted to convince people that it could be done, to prove the hypothesis.

Private donor advocacy is dependent on available financial resources, which fluctuate on the basis of factors ranging from board approval to the overall resource base of the foundation. Following the Silicon Valley dot.com bust, private foundations in the region were financially devastated. This was particularly significant for population-environment funding.[33] Overall funding allocations to these programs have declined by nearly 40% since 1997, and as a result some funders see the immediacy and popularity of attention to climate change as an opportunity to bring attention back to the population sector.[34] Regardless of the arrangements through which donors and grantees connect, the power ultimately rests with the donor—the power to create and eliminate programs, the power to support particular strategies and reject others, and, as the history of private foundations and population science suggests, the power to shape scientific knowledge production and the basis for related advocacy. For populationist donors, winning hearts and minds for population advocacy rests on the ability to convince stakeholders that population is always an imminent threat—one that is best conveyed through the scientific specter of the possible dystopian future.

Conclusion

As this chapter has demonstrated, private donors play a significant role in shaping population science from behind the scenes. I am not arguing that this is bad science; all funded scientific research is inflected in some way with the priorities of its financial supporters. However, in the particular context of population-climate science, scientific and policy-related priorities are deeply entangled, and these entanglements have a hand in shaping what is known and knowable about the future. The next chapters depart from the emphasis on knowledge-making to focus more explicitly on how policy advocates take up and circulate knowledge. Chapter 4 follows NGOs and the young people they train, exploring the everyday ways activists become advocates, experts, and development leaders of the future.

4

The Role of Youth in Population-Environment Advocacy

The previous chapter explored scientific knowledge production at the intersection of population growth and climate change, and the donors who creatively foster its growth. As noted, the research is predicated on projecting possible futures; it does not reflect the now, but the potential tomorrow, next year, or next century. It also offers a roadmap or blueprint for potential action, the kinds of actions that advocates can take to bring about what they see as the best possible future for women and for the planet. This chapter is about one such group of advocates: youth. Drawing on knowledge gleaned from the kinds of studies discussed in chapter 3, youth take up this information and carry it forward through their advocacy to influence the international family planning policy funding landscape. They also use it to produce their own futures as leaders on a global scale.

Leadership from the Ground Up

In Spring 2010, I walked the halls of Congress with a small group of college students looking for representatives to talk to about global population issues. We were in Washington, DC, for the One Voice: Reproductive Health and Population Summit, a youth advocacy workshop co-sponsored by the Sierra Club, Advocates for Youth, SIECUS, and Americans for Informed Democracy. Over the previous four days, the interactive training workshop had incorporated issue forum sessions on sex education, global SRHR, toxins in the environment, population-environment linkages (listed in the program as "sex, justice, and sustainability"), and skills-building sessions on media and messaging, community organizing, and the nuts and bolts of advocacy.

The training centered on building advocacy skills for use with congressional legislators. After learning these skills during the workshop, we would be divided into small constituency groups and sent to Capitol Hill to speak to representatives, drawing on what we had learned over the days prior. To prepare, we engaged in role-playing sessions and learned how to carefully craft messages to win sympathy from legislative audiences, drawing from both scientific data and a personal approach. In a thick packet of workshop materials, one of the handouts instructed us in the salient points of policy advocacy and lobbying: be flexible; do your homework; be prompt; be prepared; be gracious; and be professional. Using this guide and the notes we took during role plays, we sat in small groups, crafting and testing short messages that included global population growth rates and facts and figures on climate change, deforestation, species loss, and global hunger and food security; we then tied these messages to figures on contraceptive access around the world. From there, we paired up and made our pitches to members of our group, asking the pretend-policymaker in front of us to support foreign policy bills funding women's access to contraception, thereby providing a "win-win" solution that would simultaneously empower women by supporting their reproductive health and rights, while safeguarding global environmental resources and helping slow the relentless pace of climate change. We offered them handouts filled with statistics along with our winning smiles, and concluded with promises to let others in our communities know of their support for progressive issues. After several run-throughs, we were confident that we would convince at least one legislator to hear our message and consider our "ask." Off to the Hill we went.

However, our visits on the Hill did not quite follow the script. While we were scheduled to meet with four representatives, on the day of the visits two were absent, and in one office, a Legislative Aide met us at the door and asked us to simply leave the materials we had brought, thanking us for coming and promising that the representative would read them. Our facilitators had told us that they tried to match us with representatives who they knew to be sympathetic to either women's reproductive health or environmental issues, or both, but that even those who were less sympathetic to the legislative issues would offer an opportunity to practice our pitch and learn what the common responses and challenges would be.

That, of course, depended on getting into the room with a representative—which finally happened as we headed toward the lobby to leave the building, when we spotted a congresswoman known as a champion for women's issues. As she walked through the lobby, she stopped and spoke with us for a few minutes, recounting a recent trip overseas where she had learned about obstetric fistula, and expressed her horror at the practice. An outspoken member of our group—who also happened to be the youngest—seized the opportunity to share her personal passion for alternative menstrual products, and pressed literature on these products into the congresswoman's hands. This was not the plan. Not only was she off message, but we had been warned that we would have two minutes at best to make our pitch as a group, and the time was rapidly dwindling. Soon, another member of our group began to launch into the 30-second version of her population-environment pitch—referred to as the "elevator pitch," designed to be the kind of short, pithy, information-packed speech one could give in an elevator—when an aide appeared and spirited the congresswoman away. As she left, she congratulated our group on being politically engaged at such a young age (in my early thirties, I was by far the oldest member of the group), and expressed her delight at our maturity and articulate way of communicating the issues. With that, she departed with a warm smile.

We walked outside and looked at each other, deflated. Was this why we had spent the past three days furiously scribbling notes, looking at PowerPoint presentations, reading through a thick packet of handouts and brochures, and watching and participating in endless role plays? Where was our victorious moment? We never had a chance to make our policy ask. When would we have our chance to make a difference for women's SRHR?

Although it seemed to be a moment of failure at the time, what we'd experienced was in fact a demonstration of what youth-oriented population-advocacy trainings were designed to do: train and promote young people as capable leaders, able to educate and lobby a wide range of audiences on the issues that matter most to them. Of course, the issues that were supposed to matter most in this context were family planning, population growth, and the environment, issues that were at the heart of our learning and advocacy training throughout the summit. Yet, when the first trainee redirected the conversation to her pitch about menstrual products, it was the first moment in which she was able to move from a place of personal interest to broader legislative concern, and her first opportunity to translate her own personal passion into a formal policy agenda and a clear ask. In those few short moments in a congressional lobby, she transformed from an activist to a leader and policy advocate.

Population-environment advocacy serves as the basis for new leadership opportunities for young people, not just locally, but in global context. Training workshops produce knowledgeable young people who can speak with confidence and authority about global problems, using their knowledge to claim expertise and suggest development solutions. This knowledge—specialized in global problems and solutions—is a form of development expertise, and harnessing it to develop leadership skills is particularly attractive to young people who are determined to make a difference in the world as advocates.

Activism, or campaigning for social change, is different from advocacy, which is more often oriented toward directing public support to a cause or policy. The difference is subtle, but important. Advocacy requires public engagement, using one's energies toward achieving specific goals or outcomes on social issues and policies. Population-environment advocacy in particular focuses on influencing U.S. foreign policy, specifically funding appropriations for international family planning. Whether through online education campaigns, writing op-eds, lobbying members of Congress, or organizing street demonstrations, this form of advocacy takes activist dreams about safeguarding the environment and women's reproductive rights, and transforms them so that they are directed toward instrumental goals. The process for doing so requires assuming a certain type of expertise in representing global issues and problems—in this case, population growth and climate change—with authority. For the young people I studied in this project, the catalyst for that transformation, from local activist to international development advocate, expert, and leader, was the youth advocacy training.

How could this be accomplished over the course of a short weekend? One key strategy was the emphasis on youth leadership. Though led by NGOs, the summit was presented as a youth movement, one that would unleash opportunities for activists to engage more fully in social and political activism. All of the sessions in the training emphasized the importance of moving quickly from information and raising awareness to action based on youth organizing—everything from congressional lobbying actions to online awareness campaigns—and youth ownership of the future. As one of the speakers told the crowd during the opening plenary session, "The older generations need us; they need to ask for our input. The global movement is turning international development on its head, including those most affected by it in the planning process. This is our civil rights movement."

Over a period of nearly two years, I participated in a series of trainings and workshops organized by the Sierra Club's Global Population and Environment Program[1] and their partners: various organizations that were active in some way on population, family planning, and reproductive health. These organizations included Planned Parenthood, Advocates for Youth, the Feminist Majority Foundation, Population Action International (PAI), SIECUS, and the International Women's Health Coalition. During that time, I explored the ideas underlying the notion that young Americans could represent the interests of women at vast geographic, cultural, and economic distances, that these interests could be encapsulated by a focus on contraceptives, and that this advocacy was motivated by feminist desire to empower women. I also investigated how this advocacy was framed, both by new advocates and those who trained them, as being tightly linked with notions of social justice.

My Own Location

But first, a personal note. This is a challenging chapter for me to write, precisely because of my own ambivalence toward the trainings, the messages they disseminate, and how those messages land with new advocates. When I conducted this research, I was several years into a PhD program, was well versed in the history of population control, and had long since sworn my allegiance to political ecology frameworks (see the introduction) for an understanding of the causes of environmental problems. However, I had also worked in international public health organizations for several years, and I understood the lure of wanting to take concrete action to impact global problems. Just a decide prior, as a young and energetic idealist, I joined the Peace Corps, where I served in rural Madagascar as a Community Health Advisor. There I spent my days talking to young people about sex and reproduction, life skills, and how to negotiate condom use. I also taught conversational English, usually on the sexual and reproductive health topics I was so passionate about. I loved my work, even if I spent many homesick days longing for the familiarity and comfort of the U.S. But it wasn't just the work itself that I loved; it was the sense that I was doing something that could make a difference, though exactly what kind of difference was rather murky. The frustration of my undergrad days of learning about social injustice, deep poverty, and gender inequality across the Global South, and Africa in particular, finally had an outlet. I was taking action.

Malkki's study of Finnish international aid workers is illuminating in this context. She found that people who are motivated to "help" by becoming involved with international causes are not necessarily cosmopolitan, global, or worldly people. Rather, their impulse to help unknown others, thousands of miles away, is motivated in part by a desire to escape the familiar. In this way, the "international humanitarian imagination"[2] is actually quite domestic in that it begins at home, in the local, with local histories and political conditions, and with people's conceptions of themselves as members of their own societies. This imagination constructs vulnerable others as always already in need—of sympathy, of aid, of intervention. But what of the other need, the "co-present neediness on the other side, *the neediness of the helper, the giver*"?[3] Insisting on the neediness of the humanitarian would-be benefactor locates their social and existential position, bringing them out of the shadows of the unmarked benevolent actor. This project contributes to that inquiry, interrogating the motivations and desires of today's population advocates-from-afar.

In my own case, I realized that my disorganized actions in Peace Corps were based on vague notions of right and wrong, and predicated on the idea that, because I had a graduate degree in Public Health, I possessed expert knowledge that the people around me did not have. My Malagasy counterparts and friends let me know in subtle and not-so-subtle ways that my knowledge of statistical data and health campaigns that had worked in other settings were of little use to them. They did not fit the contours of their lives. Eventually, I took a step back, began to listen, and recognized that I had much more to learn from Malagasy girls and women than I could teach them.

All of this is to say that I understand what motivates young population-environment advocates. The desire to do good work and make a difference in the world, particularly in conditions of extreme inequality, is a noble ideal, even as it is often informed by vaguely colonial notions of saviorhood. I felt an affinity for the people who conducted and participated in these trainings; I sensed that, had I been twenty-two years old and not yet a Peace Corps volunteer, I might have become a population advocacy trainee too. With that said, given the path I did take, and which ultimately led me to this project, I was fairly skeptical about this work and what it sought to do. Were these young people familiar with the history of population control in the U.S. and globally? Had they studied the Malthusian-driven science that justified and helped make population control possible? What made them think they could represent the interests of women and girls they had never met, whose life conditions were completely foreign to them? Did they actually understand what, and who, they were advocating for? And finally—did the quest for leadership and action outweigh other concerns? In the opening scene of this book, I described a student art reception linking population and environment issues for a youth audience. Here, I return to it to illuminate how youth leadership and advocacy are built, with NGOs operating from behind the scenes.

Advocacy Is Fun!

The invitational flyer announced the "Sex and Sustainability" event as an opportunity for UC Berkeley students to come together for food, drink, art, and ideas about global sustainability, women's rights, and justice. I arrived to find it in full swing, with over 200 participants in attendance. The walls were filled with artwork on gender and sustainable development that had been commissioned for the reception. Peering at the fourteen mixed media installations scattered throughout the gallery revealed images of faceless women bent double, babies strapped to their backs while they toiled in the fields; women's bodies engulfed in swelling pregnant bellies; and an emaciated child encircled in a prison of skeletal bones. The images were all painted in the beige, yellow, and deep brown paint pigments selected to represent racially recognizable darker skin tones. The caption beneath one painting read, "Let's see who dies this time, me or my baby."[4] Each image was also accompanied by statistics referring to fertility rates (the number of babies the average woman in a particular country will have in her lifetime), maternal death rates, childhood illness and death, and hunger. They set a rather morbid tone.

Walking around the gallery, I overheard a few snippets of conversation; most students were not discussing the artwork, but rather their classes and spring break plans. A few snickered at a professor wearing a tie emblazoned with condoms. As I stood looking at the artwork, another graduate student I knew approached me, aghast. "What the hell is this?!" she asked, waving at the paintings. "It's like, you don't need to offer any explanation. Look at the poor darkies, look how depressing life in Africa is." She rolled her eyes, finishing her wine and refilling her cup. I vaguely agreed with her, adding that for most of the students in the room, this was probably their only entry point into thinking about these issues. We went back and forth for a few minutes about whether images that present essentialized ideas of women, motherhood, and poverty are ever useful. While we talked, a young white woman who worked for the organization co-sponsoring the event overhead our conversation and interjected. "It sounds like you have strong feelings about this. You should probably say something to the organizers about it." We regarded her silently for a moment, then I asked, "Well, what are *your* feelings about it?" She gave me a wry smile, noting that she agreed with us that the images were simplistic, but said that they work. "We're just trying to get people in the door here," she noted. "After that we can offer more nuance."

In her work on the history of gender and environmental policy work at UN environment conferences, Resurreccion demonstrates that essentialism played a significant role when feminists in the 1980s were trying to bring women's issues to the table. Presenting women's relationships to nature in the Global South in homogenous ways provided policymakers with a shorthand for understanding the basics of gendered environmental roles and differentiated impacts. At the same time, this practice gave rise to persistent storylines linking poor, rural women to the environment as both victims of environmental catastrophes and also agents of change, responsible for solving environmental problems. These narratives fix women in a static model: that of an essentialized, universal relationship between women and nature. As a result, the initial policy attention to women's environmental experiences left little room for articulating dynamic, diverse experiences in ways that resist the victim-agent role.[5]

The artwork in the gallery was reminiscent of this shorthand, and reflected other images that are common to population advocacy. The trope of the singular, unchanging Global South woman appears frequently in images and texts found in population-related brochures and reports. This image has changed to the extent that the referents used to describe her have shifted, from "Third World woman" to "Global South woman," "Impoverished woman," and "Woman in developing countries"; however, she remains a flat, static image comprised of several characteristics: poverty, high (above replacement) fertility, and significant barriers to limiting her childbearing. Whether in Ghana, India, Madagascar, or Cambodia, this singular trope endures, shaping language used to describe poor women, the conditions and causes of their fertility and childbearing practices, and why family planning advocates in the U.S. should care to advocate on their behalf.

Marisa, then-coordinator of the GPEP, is a cheerleader for population-environment advocacy. She has an exuberant, infectious smile and charming demeanor, and the passion she projects for her work is hard to resist. It is clear that she cares deeply about women's reproductive health. In her early 20s, white, and middle class, she swirled effortlessly through the crowd of other mostly white, young women students wearing a GPEP t-shirt emblazoned with the slogan "The fate of the world is in your hands . . . and in your pants." I asked her what attracted her to this work, and she replied that she had majored in Sociology in college, and had always been involved in feminist issues: "I've always been attracted to diversity and feminism. I think this is a really good way to make a difference for women's rights," she noted. Another GPEP coordinator, Leslie, also a white woman in her early 20s, was also deeply committed to population-environment advocacy. For her, it was common sense: "Increasing human numbers are having a huge impact on our world, in every country, not just in the Global South. The link is simple. More mouths to feed means more food to be produced and consumed, more land being farmed, more emissions being generated. But we also have to think about poor people and how they're impacted. The linkage goes both ways." Leslie had read *The Population Bomb* in college, and initially approached the GPEP with some trepidation. "When I first heard about the program, I thought controversy, right away. I knew about the history of population control. But I did some research on what the program really stood for, and I'm really passionate about climate change, women's issues, and women's health. For me, I'd never seen any other organizations that were actually doing this work and addressing the linkages."

Several months later, I encountered Marisa again at a multi-day GPEP advocacy training. It was held in Beverly Hills, California, at the Feminist Majority Foundation (FMF) headquarters, and both FMF and Planned Parenthood were co-sponsors. Before the meeting started, as I joined a few women milling around the breakfast table, stirring our coffee and picking over donuts and bagels, I noted the demographics of the crowd: approximately 30 women, mostly college age, mostly white, although several women of color and a small handful of post-retirement-age women were also there. We sat at tables in the conference room and opened thick folders filled with the workshop agenda, biographies on our facilitators, and articles ("Dying in the Backstreets" chronicled the practice of unsafe abortions in Kenya), fact sheets (on the U.S. government's woefully inadequate funding for international family planning), and information sheets exposing fake women's health centers operating near college campuses. The folders also included swag: buttons from Planned Parenthood ("Love Carefully") and the Sierra Club ("Pro-Choice; Pro-Family Planning; Pro-Environment"), as well as green and pink Sierra Club condoms and stickers, and t-shirts with their youth slogan.

Our resource packets told us that by the end of the workshop, we would all be able to do the following: 1. Describe the relationships among global-to-local issues such as sexual and reproductive health and rights, resource consumption, global warming, population, poverty, and gender equity; and 2. Understand the importance of voluntary family planning and sexual health education here and abroad. The agenda broke down the schedule for the two-day training in thirty-minute and hour-long increments, including sessions on "Global Reproductive Health and the Environment," "From Global to Local: Family Planning and Sex Education," values clarification, skills building and role-playing exercises, and sessions on action planning and community outreach. The interplay between information sharing and action quickly emerged as an important component of the agenda; it was clear that although this training would share knowledge, the emphasis would be on building action-oriented advocacy skills. By the end of the training, we were expected to create action plans that would "make a difference on your campus and/or in your community (for example, by hosting an educational event or film showing, contacting a decision-maker, recruiting other volunteers to get involved with an existing campaign, etc.)."

Taking my seat, I asked the young woman next to me what brought her there. She told me that she was primarily concerned about environmental pollution, loss of wildlife habitats, and environmental justice. She was an avid hiker and wilderness enthusiast, and felt that people were generally careless about their environmental practices. "Look at the packaging we use every day," she noted. "These water bottles will be on the earth forever. Forever!" I agreed, and asked how she saw the connections with reproductive health and family planning in the Global South. She hesitated for a moment, then said, "Well, I've never been outside of the U.S., so I don't really know. But if it helps women have access to health care, it's a good idea."

I thought about her statement as the first session got underway. Did we need to have global experiences or perspectives to advocate for global population issues? What was required to be a global actor? The answer, apparently, was data. A session facilitator launched into a dizzying array of statistical data: global population numbers (6.8 billion at the time), the number of women who wanted to stop or pause their childbearing but weren't using contraceptives (over 200 million), the number of women dying in childbirth (one per minute, more than half a million each year), and where these women were located (99% in the Global South). The solutions to these problems, she claimed, were simple and technological: contraceptives, safe motherhood kits, and safe abortions. Education was presented as a simple, technical intervention too: "What's the number one factor determining how many children a woman will have in her lifetime?" the facilitator asked, looking around expectantly. Several participants in the room were FMF interns, and they mumbled the word "education," to which she shouted, "Exactly!" Education, she told us, is a crucial component to protecting women's health, preserving lives, and empowering our Global South sisters. With education, the shocking statistics we'd heard at length would no longer accurately represent the health status and life chance scenarios of women around the world.

The room was comprised of a group of relative elites (middle-class, college-educated people) grappling with how to understand and represent the life conditions of people who live very far away. This distance is both literal and figurative. They could not fully wrap their heads around the concepts of child brides, female genital cutting, and obstetric fistula: "They all sound really bad," one participant noted during an open discussion period. "Women really need education." And yet, the facilitators never discussed the more complex, nuanced ways in which poverty, limited political power, and cultural norms also shape women's reproductive lives. They also never discussed the importance of meeting basic needs or having access to the tools of collective organizing for a voice in political decision-making. Why? Did they deem it irrelevant, or too complex for the audience? Or did it simply not meet the needs of the program, to produce new advocates?

These questions continued to occupy my mind throughout subsequent training workshops, gaining in complexity as the groups of attendees around me became more diverse in background and interest, and it became clear that they were not entirely sure of their own motivations and desires for population advocacy. Some tension around these questions bubbled over the following year at the Youth Summit in Washington, DC—the event in which youth participants trained for lobbying visits on Capitol Hill. Day 2 of the summit included a session on "Sex, Justice, and Sustainability" that honed in on population and climate change as a unique way for activists to amplify their voices on a global stage. Organized and led by GPEP and PAI, the session focused on the central message that global reproductive health and family planning for women are essential ingredients in promoting environmental sustainability and advancing climate change mitigation and adaptation. Representatives of the two NGOs used colorful slides filled with charts and statistics on GGEs, population growth projections, and contraceptive access in global context, highlighting the role of advocacy and raising awareness in linking the three. Their bottom line was that population, climate change, and contraceptive access were not distinct issues, nor separate from politics and policy. Rather, what joined them together was the necessity for activism and policy advocacy to draw the links to global social justice and women's empowerment. At the end of the presentations, one of the panelists added, "It's such a beautiful story. Giving women what they need and preserving the environment. It just comes together in such a beautiful way."

Looking around the room, I noticed mixed reactions to the presentation. Of the two dozen participants, most of them young women in college, approximately half were women of color. Of the women of color, nearly all were attending the summit to promote sexual and reproductive health, rights, and justice among marginalized American populations. Their perspectives and priorities were local, not global, and the strategies, data points, and languages they were familiar with reflected that. The data on climate change and environmental sustainability seemed disconnected from the contexts of contraceptive and abortion access in low-income communities they discussed in small groups. Many of the young white women in the room were more concerned with environmental problems, identifying as environmentalists or climate change activists. The global preceded the local in importance; their language was peppered with references to climate change as "everyone's problem," an imminent danger for "all of our lives."[6] As one young white woman stated, "I've been an environmentalist my whole life, I consider myself to be an activist, up on all of the most important issues, but I'm embarrassed to say I've never even thought about reproductive rights. This is completely new to me."

These differences became particularly salient during the post-session Q&A, when a white participant stated that "overpopulation" was a key threat to environmental sustainability. Others in the room responded swiftly: panelists and experienced participants encouraged new trainees to remove terms like overpopulation from their vocabularies in favor of "population pressure," "unsustainable growth," and "rapid population growth." Meanwhile, participants of color asked who was considered to be overpopulating the earth, and which populations would be targeted for reductions. The mood in the room became uncomfortable as facilitators attempted to navigate the linguistic and social fault lines that had opened up, smoothing the moment over by drawing everyone's attention back to the central goal of contraceptive access for all women around the world. Regardless of whether we were primarily concerned about environmental problems or women's reproductive autonomy, they reminded us, full and voluntary contraceptive access was something that we could—and should—all be on board with, as a starting point for justice across social arenas. Moments like this would repeat through numerous training workshops I attended, and while at first they struck me as moments of skillful language policing, it quickly became clear that something else was at work. These were moments when population actors were reshaping the language in which their efforts were understood, and they were repositioning it—moving it away from the thorniness of controversy, of coercion and human rights abuses and questions about racial inequality. They were reframing population as a social justice issue, and in doing so, they linked their efforts to a similar group of populationist youth who were active several decades prior. Ironically, that group was an explicitly self-styled population control organization, but its leaders navigated complex relationships with women's rights activists, family planners, and neo-Malthusians. That organization was known as Zero Population Growth (ZPG), and the movement it created provided the model for youth population advocacy today.

ZPG and the Roots of Youth Advocacy

One day in 1968, students at Yale University awoke to find the campus papered with flyers and stickers bearing a rather cryptic message: the letters ZPG. There was no explanation of what they stood for, and the ubiquitous and unexplained presence of the materials generated a buzz that spread rapidly throughout the campus community. Two weeks later, the mysterious posts were replaced with different materials, this time explaining that ZPG stood for Zero Population Growth, a new club on campus, and advertising their first meeting. Within a week the group boasted a membership of 200 students, and over the following year and a half, Yale became one of the leading campus chapters of the club.[7] Meanwhile, ZPG was growing rapidly nationwide: in the two years from the time of its founding in 1968 to 1970, the organization grew to 102 chapters in thirty states, and was adding five hundred new members each week, many of whom were college students and other youth. By that time, ZPG had also hired a Washington, DC, lobbyist, and—in California—its leaders initiated efforts to set up a state commission on population and environment.[8] ZPG had gone from a young people's social movement to a formal NGO (it continues in this capacity today, under the name Population Connection).

Both the organization's message and its strategies—using bumper stickers, buttons, and flyers with catchy phrases and slogans—were successful in drawing attention from large numbers of people in a short period of time. They offered young people a strategy for something they could do, right here, right now, that was entirely under their control. ZPG aimed its messages primarily at the white middle class, encouraging them to participate in global stewardship through responsible management of their own reproduction. This approach fit well with feminist advocacy for abortion access and legalization of contraception, and in the context of the Free Love movement, the sexual revolution, and the women's reproductive rights movements, it was a time in which talking openly about sex and gender was becoming increasingly commonplace. ZPG was concerned primarily with bringing down birth rates in the U.S., particularly among the white middle class, by restricting their own birthrates to two children or fewer, connected to outsized global resource use and environmental impacts. Supporting abortion access, legalized contraception, getting rid of tax exemptions for children, and welfare reform in the U.S. were also key goals.[9]

At the same time, ZPG leaders were split on the issue of voluntarism versus coercion. Garrett Hardin, an advisor of the group, was a prominent advocate of coercion, and Richard Bowers, one of ZPG's co-founders (Paul Ehrlich was another co-founder), argued that "voluntarism is a farce."[10] Meanwhile, ZPG Executive Director Shirley Radl testified before the California legislature in 1970 that one of ZPG's central goals was to create the conditions for all members of society to voluntarily limit family size, arguing that all forms of contraception and abortion were necessary to achieve this goal. Underscoring her point, she declared that "the uterus should be the concern of the owner and not the state."[11]

Radl was not the only activist involved in both the growing environmental movement and the mainstream women's liberation movement, viewing the two struggles as natural allies. Lucinda Cisler, the Chair of the National Organization for Women's Taskforce on Reproduction, was also on the national board of ZPG. Feminist activists who sought to delink womanhood from motherhood, arguing that women had important contributions to make to the world far beyond childbearing, found ready reception among environmentalists who also argued that women should have a greater choice of lifestyles and options available to them. Garrett Hardin pushed strongly for the complete repeal of abortion restrictions, and helped form the organization National Association for the Repeal of Abortion Laws (NARAL).[12] At the same time, ZPG leaders lacked nuance in terms of how the issues unfurled—particularly the issue of sterilization. In their zeal for full access to all forms of contraception and abortion, they opposed all legislation limiting access to sterilization, going so far as to join in a lawsuit with the ACLU on behalf of a group of white women who pressed for expanded sterilization access. As the next chapter demonstrates, this was devastating to women of color activists, who were seeking greater restrictions on the procedure, which had been used coercively in their communities for decades.

The organization also called for land use planning designed to "determine the appropriate patterns of distribution of people on the land, and of migration between states and regions,"[13] and overall stabilization in national energy consumption, as well as immediate reduction in the consumption of non-renewable energy sources. They saw this as necessarily concurrent with the achievement of improved access to health care, education, and increased income levels for all. While they did not use the words, their approach was rooted in what today's young advocates might describe as social justice.

The growth of ZPG highlights a key contrast within population-environment activism that has endured until today. While young people—particularly college students—formed the bulk of its membership, it was not only a campus-based club, nor were its founders and leaders "youth." Paul Ehrlich, Charles Remington, and Richard Bowers (two professor-scientists and a lawyer) were well into their professional careers when they founded ZPG, and the organization became a formal nonprofit organization, not just a student movement. Today, population-environment activism is similar: while many of the voices of population advocacy are youth voices, those leading youth-organizing efforts are employees of organizations like the Sierra Club.

The Sierra Club is not alone in focusing on youth for their population-environment efforts. When the Center for Biological Diversity (CBD) in Arizona launched its Endangered Species Condoms campaign in 2010, they focused one of their early distribution events at the University of Arizona. They produced 100,000 condoms for their Earth Day launch, which turned out to be a gross underestimation of demand. According to the then-program director, student response to the campaign was overwhelming: expecting to distribute hundreds of condoms at the university, he ended up disseminating thousands, as students waited in long lines for hours. The campaign produces and distributes condoms in boxes bearing punny slogans,[14] and when opened, users find the inside of the box printed with information on endangered animal and insect species, juxtaposed with statistics on population growth in the U.S. and messages urging the condom holder to "save panthers, sea turtles, wolves, and countless other endangered species by choosing to stop overpopulating the planet." I've shared these condoms with my students in my own classes in hopes of sparking a critical discussion of the ways environmentalists frame population issues, but without fail, any sense of critical analysis gives way to entertainment. Students compare packages, laughing at the witty phrasing, and describe the discussions they have with their partners when negotiating condom use (which, to date, have never included saving beetles, jaguars, or any of the other animal and insect species found on Endangered Species Condoms). While the packages are successful at reviving students' flagging attention during class, they don't necessarily inspire serious thought about the relationships between population growth, contraceptive use, and the environment.

Nonetheless, they were wildly popular with young people from the very beginning. CBD distributes the condoms for free through volunteer distributors, and while the campaign has never been limited to a youth audience, young people quickly emerged as a key constituency that was interested and motivated to distribute condoms to others. The former program director told me that students often wrote to say that the condoms were useful to their awareness-raising campaigns on campus. As a result, when students would request condoms to distribute, he would send up to ten times the number they requested.

The Sierra Club and its partners have found a way to harness this youthful energy and attraction to sexy language: by invoking social justice ideals. A standard message throughout training workshops held that framing population-environment connections through women's reproductive health, rights, and empowerment had never been done before, and that previous approaches—focused on population control—were a part of the "dark past,"[15] never to return. From there, the focal point of trainings, the strategy that would firmly move population from the darkness into light, in their minds, was clear: populationism as a form of social justice.

(Re)Framing Social Justice

In 2010, *A Pivotal Moment: Population, Justice, and The Environmental Challenge*[16] outlined the new concept of population justice. The book is an edited volume that describes the state of the world as occupying a pivotal moment both environmentally and demographically, driven in part by "the largest generation of young people in human history"[17] coming of age and holding the reins of the future in their hands through their decisions about sexuality and childbearing. The book offers a solution: population justice. Described as an ethical framework for addressing population and the environment, population justice draws on reproductive and environmental justice to contextualize reproductive and environmental behaviors and constraints, with a "nuanced understanding of the relationship between human numbers and environmental harm."[18] Population justice is rooted in the Cairo approach to women's human rights and bodily integrity. It is also, notably, committed to the Cairo Consensus's emphasis on individual responsibility: "If our basic rights are secured (a big 'if' for many people in the world), then we have an obligation to ask what impact our choices have on others, including future generations. In the context of an unfolding environmental crisis, the question is an urgent one."[19]

The book situates population growth and environmental change within broader political and economic conditions, and it advocates access to health care, education, and sustainable development; at the same time, the solutions it offers all come back, in one way or another, to family planning. Moreover, it depicts contraceptive access as a social solution, rather than a technological one, arguing that "if our goal is to create a world that is sustainable and just, population-environment policies must serve those ends."[20] How this notion of justice is defined is rather vague; it is up to the reader to interpret its meaning (note that several chapters in the edited volume focus on addressing human numbers in arguably neo-Malthusian ways). Yet, the ideas behind the book served as the inspiration for a central pillar of the GPEP program: its youth leadership program, known as the Population Justice Environment (PJE) Challenge. In an interview with Marisa, she argued that GPEP understood the justice framework from the book in strategic terms, particularly for GPEP recruitment efforts:

JS: Why focus on justice? What do SRHR and justice have to do with global population and environment?

MARISA: We've got our program into a place where the message is really good. We found *A Pivotal Moment* and went to the first book strategy meeting in January of 2008 and I think that's really where the idea sort of sparked around what does population justice really mean, and how does sort of adding the word justice not just sort of make people feel comfortable with these issues, but also to help them question their values, on these complicated issues. We brought it in strategically at every stage. We had her speak at our One Voice summit last year, you know, and she really framed the whole conference in that frame of justice and equity, and I think the youth really appreciated it and appreciated the issues. And the millennium generation is really craving synergies. They see how these issues overlap, and that sort of messaging goes really well into bringing issues together and also uniting environmental groups with women's groups, women's health and rights groups, which environmentalists haven't always had the best relationships with.

This notion of social justice grounds the concept of youth leadership within GPEP. It coheres around a concept of self-appointed responsibility to take action to turn the tide of population. Unlike the ZPG programs it indirectly emerged from, this form of youth leadership moves beyond a duty toward reproductive self-management, and simultaneously positions young advocates to argue for Global South others to do the same. At the same time, it fits with a narrative among population-environment advocates that they are doing the work of social justice. However, their own descriptions of social justice and how it is embodied in their advocacy work were vague at best.

This aligns with what seems to be a rising trend among young people nationally. Among the many things I've learned from my students in the classroom over the past ten years, one thing is abidingly clear: they were, and are, deeply interested in social justice. This frequently comes up in class discussions, most often in reference to national politics, but also pertaining to national and local social movements. Our classroom discussions revolve around a diverse range of issues and topics: police brutality, Black Lives Matter, food deserts, rape culture, immigration politics, the Occupy Movement, climate change, and toxins in food, air, water, and cosmetics. In each case, more often than not, the conversation would come back to social justice; not how it is defined or what it means in practical terms, but rather how to achieve it based on taken-for-granted assumptions of a shared starting point and common values around equality and rejection of "oppression."

Through many of these conversations, and particularly in the context of studying young activists, I came to realize that the concept of social justice was losing its significance, as it was buried under an avalanche of assumptions of the "right" causes to rally around. As Reisch notes, moralizing language is never far behind, in ways that make it ever more challenging to parse out the meanings of social justice and its adherents: "As desirable social and political goals are depicted in starkly different forms, labels like 'good' and 'evil' become interchangeable and the meaning of social justice becomes obscured . . . If liberals and conservatives, religious fundamentalists, and radical secularists all regard their causes as socially just, how can we develop a common meaning of the term?"[21]

This watering down of the concept helps youth-oriented programs like GPEP draw in activists with simplistic messaging focused on action. GPEP program documents argue that engaging youth activists is a necessary strategy for changing U.S. policies that impact youth around the world, and that "young people must have a voice and a seat at the decision-making table when it comes to ensuring the health of their own bodies, and the planet!" Marisa reflected some of this approach during an interview, arguing that the issues GPEP advocated would primarily affect youth in the long run, and therefore it made sense to support and foster the leadership potential of youth during training efforts. After all, "half the world's population hasn't entered their reproductive health years yet . . . and given the political environment today, we have engaged youth and it would be a pity not to maximize their potential," she stated. GPEP also recognized that they had a prime opportunity to draw on young people's access to technology and social media to disseminate messages and mobilize advocacy efforts quickly. Another GPEP employee agreed: "Classes and the web make information available in ways that it hasn't been before. The world is so different from when I was a college student. Technology has advanced rapidly, and the approach is very different . . . I think technology has an important role; we're getting exposure, and the media is so different than it used to be. In the digital age, in some regards, people get exposed to things differently. When we talk about water being an issue, they [youth] can immediately go and look up images. So we have to draw on their strengths and maximize them."

Through interviews and training activities, youth advocates expressed a desire to make a difference on a global scale, specifically by improving the lives and health of others and the planet, based on social justice ideals. And while the arguments linking population and the environment may have been new for some, they resonated deeply because they are, as I was told time and time again, a matter of common sense. When I asked one young woman, a campus environmental justice activist, why she had become involved in population advocacy, she responded, "It just made sense. I want to work to create solutions that affect everyone based on justice. All people are affected by sex and sexuality, and all people are affected by the environment . . . Also, women work to get natural resources their families depend on. I think that choosing how they want to have and raise families is one of the things that can most improve women's status in a society. It can't come from outside; we need to support women to do it locally within their own societies. And if women have less children, it benefits all of us."

A former manager at the Sierra Club, who has had a long career in population-environment advocacy work, explained that meeting youth where they are and building on the issues that are most interesting to them is a key part of the organization's strategic approach to engaging young activists. "Within the Sierra Club, youth activists self-define not as environmental activists, but as climate change activists, and are interested in justice and a rights-based approach," he noted.

But this focus had its frustrations, particularly for trainees. While Sierra Club youth trainees may have seen climate change as a main priority, their interests were not necessarily shared among their friends, particularly when it came to population. A long-time youth population advocate expressed frustration with the challenges of getting other young people on board: "It's a hard issue for a lot of people to care about and understand. In the environmental world, it's not a sexy issue, it's one that people want to shy away from, especially environmental groups that don't want to get involved with population control. With young people,

I don't see much growth on the issue. They'll say they're interested but they end up working on other issues instead. Within climate activism, there are trends and people end up working on one issue or another but this one isn't sexy, even though it's about sex."

At the same time, other youth advocates saw more opportunities to connect population-environment issues to other arenas of life, where the audiences they wanted to reach were more likely to be concerned. One South Asian graduate student I interviewed approached the issue through a public health lens, based on her reproductive health background, and had attended a Sierra Club event linking the issues to the environment. That event marked the first time that she ever connected environmental issues to reproduction: "So my focus on environment came from Sierra Club, when I was interested in population issues. They connected it to environment, I didn't. And they gave me money to do an advocacy event based around population and environment stuff. And, so when I went to their training, they introduced all the connectivity between the environmental issues and the population issues, which were sort of in the back of my mind I guess, but they brought them out into the forefront."

She also expressed ambivalence and concern about exactly how the linkages are made, but argued that it was necessary to frame population in new ways to attract broader support for family planning. At one point during our interview, she confided that at times she was hesitant to talk about population and environment because there were more compelling issues to discuss than the environment, but that focusing on the environment created more opportunities to talk about reproduction: "Generally I think that the reason why population, rapid population growth is coming up as an issue is because it's connected with the environment. It wouldn't be coming up so much if it wasn't for the environment. I think environmental issues are pulling it up out of the dredges of things that you can't talk about."

Another interview with Marisa revealed that intense polarization on population offers deep challenges to framing and circulating messages, regardless of the politics of the listener:

JS: Are population-environment linkages still controversial today? Or has that changed?

MARISA: My friend has a great quote which is, "Population is one of those issues where you're attacked from the left and from the right" [laughs], which I think sums it up so well. It's really polarized. The further you go out to the poles whether it's women's health and rights, resource distribution and conservation, the pope, the demographic lens, the further you go out to those poles, the more controversial the issues become. But in the middle, broadly speaking, if you talk to Joe Schmoe on the street, I don't think that these issues are that controversial. You know, I think that generally people acknowledge that yes, there's a connection, and yes, we live on a finite planet, and yes we should support women's health and rights. Just a simple set of solutions that are really important to support, but I think the international coalition on family planning did a poll that said that something like 80% of people support international family planning. And I really want the PJE Challenge to have that basic support because I think these issues become heated based on the audience.

JS: Who takes issue with it? Are there certain groups who are more likely to be opposed to this kind of work?

MARISA: It's both the left and the right. I was attacked two years ago by Fox News; I was giving Sex and Environment presentations, and they did this big article saying sex can also cause global warming. It was hilarious. That you would expect from conservative groups who aren't about foreign aid or family planning. Then there's also these feminist groups that just sort of flat out believe that environmental organizations have no claim in talking about population. They say it's just an issue about women's health and rights, and gender equality, and just . . . there should be no connection between women's fertility and environmental resource use. That's definitely an attack from the left who are so pro human rights to the extent that they ignore the biological fact that we live on a planet with finite resources, and also looking at any country that has escaped poverty has slowed their population growth. It's almost an argument that's too academic, or too rooted in trying to be progressive.

A former GPEP employee, Nicole, saw the main issue as one of developing strategic communication and build effectiveness while minimizing controversy, and it quickly became clear to her during the early GPEP days that situating population as an international problem outside of the U.S. was a way to do so. This is in part due to the fact that domestic population issues are linked to immigration and population growth rates among different racial and ethnic groups. Given the Sierra Club's own history of controversy on this issue,[22] it became clear that avoiding the issue played a major role in framing their population work's Global South focus:

JS: Why focus on population issues overseas?

NICOLE: Sierra Club has addressed population growth since the '70s, and has had a history of controversy around the issue. There are segments of the membership that feel that population is the most pressing environmental issue, but if they want to break it down into finding solutions, it's "yes, let's increase access to services, but let's also decrease the number of people on the planet, and let's close our borders." For an organization like Sierra Club that's really trying, like I'd say big, white environmental organizations, having a program that works on immigration reform, it's just gonna burn a lot of bridges. And for the many reasons why Sierra Club doesn't work on it politically, it would make sense for the organization to address it on many levels. Throughout my history there it was always an issue and we had to be very clear in how we spoke to the public and the opposition. It's always going to come up. When talking about demographics or population issues, immigration is a part of the challenge.

JS: What about the argument that says we must address US population growth first, since we're a leading emitter of greenhouse gases?

NICOLE: This is an area where justice and human rights have to come first. It's not about control or limiting choices or implementing culturally insensitive policies. I was very aware that you had to articulate these issues from a point of focusing on equality. I think the climate change issue is a slippery slope. We at GPEP would talk about how a growing population impacts the ways we use resources and etcetera. And of course we couldn't talk only about population growth in the developing world when we're the biggest emitters and polluters. But ask me where the sound bite or talking point is, and I couldn't tell you.

JS: Is there a correct way to talk about population and climate change?

> NICOLE: It's very challenging on climate change. Sierra Club is the one environmental organization right now that is tackling population issues. Is family planning the solution? Absolutely not. It's a piece of the puzzle when looking at the broader issue of looking at issues of sustainability. Wealthy nations all play a role in a way that we consume resources. But if we look collectively at the world, since there are no boundaries with climate change, we still need to look at realistic growth. That's not the cause, but in looking at the future of sustainability, it has to be a part of the conversation.

Again, youth population advocates have multiple objectives, leadership and taking action primary among them. Minimizing controversy is seen as an important way to push past the difficult and complex issues youth advocates see as a barrier to accomplishing their goals—and situating their work as globally necessary often emerges as a useful way to do it. For example, when a small group of youth advocates, several of them GPEP trainees, drafted an updated policy document on population growth, family planning, and climate change for an international climate change conference, they foregrounded their own sense of responsibility and leadership as leaders. They also conveyed a sense of global connectedness to other young people around the world.

The policy piece, *COP16 Policy Statement—Global Youth Support Sexual and Reproductive Health and Rights (SRHR) for a Just and Sustainable World*[23], is a comprehensive document that sums up SRHR and climate change through a central framework of collective youth leadership. It opens with a statement of the unique role young people can play in addressing climate change: "Collectively, we as young people have a critical role to play in adapting to climate change, helping mitigate climate change, holding our governments accountable to targets set in Cancun, and shaping a just and sustainable world." From there, it addresses the gendered livelihood impacts of climate change—the intensification of women's roles with respect to agriculture, fuel and water collection, as well as the disproportionate gendered impacts of natural disasters—alongside the reproductive health challenges facing young women on a global scale, including the unmet need for contraceptives, complications of early pregnancy, and overall low access to quality reproductive health services. Grounding these problems in poor political leadership and young people's disenfranchisement, the policy statement describes young people, alongside political leaders, as uniquely positioned to intervene and transform global conditions through their own actions: "There is growing evidence that addressing SRHR solutions can increase resilience to climate change and slow population trends that exacerbate poverty and climate change impacts, empower young people to exercise their rights and contribute to achieving a more just world. Governments have committed to delivering these services through a number of international agreements. What we need now is action."

This language communicates two key ideas: first, that young people bear unique climate and reproductive health burdens due to woefully inadequate leadership from older generations, and second, that they are the right people to solve these problems through their advocacy. The remainder of the document lists seven areas of government action, which it advocates strongly for world governments, particularly UNFCCC member state leaders, to take. These include: funding and supporting the least developed countries' climate change plans, investing in youth-friendly SRH education and services, ensuring girls' and women's access to education worldwide, and providing universal, voluntary access to family planning, with young people participating fully in program development and planning.

I spoke with one of the architects of the statement, Julie, a GPEP training alumnus and outspoken advocate on population and environment issues. She indicated that she and other advocates were motivated to draft the document based on a sense of frustration with the limitations of older activists' thinking. Julie argued that the SHRH community has come late to the climate change table, and that young activists were needed to fill the void; thus, they wrote the statement to show others how to integrate climate change within existing SHRH advocacy. When they went to the COP15 meetings in Copenhagen and circulated an earlier version of the statement among gender and health activists, they were initially met with pushback, particularly from older feminists in the daily gender caucus meetings. Julie recounted one particularly contentious encounter: "We got really strong emotions generated from older feminists when raising the issue, before they read the statement. The gender caucus included people from all over the world; those expressing concern were from both developed and developing countries. There was a strong passionate response of, 'we don't want to impose programs, tell women what to do, or blame them.' It was useful for me to realize that people with those backgrounds . . . you have to start by acknowledging the bad part, show that we know the history and talk about the approach now."

Despite the tension, Julie and her partners counted it a success because they had taken the opportunity to assert leadership and to push for what they saw as an innovative approach. At the COP meetings, youth voices are recognized as the future of climate activism, so much so that young people operate their own informal conference just prior to the larger, official proceedings. Asserting leadership by circulating and gaining support for the policy documents, sharing the message with older activists, and adopting an authoritative tone as leaders of a global youth movement all served to bolster their own sense of being global actors who can move forward international policies where older people lag behind. Although organizations like the Sierra Club play a key role in the development of their ideas, they fully assert that they are at the reins of pushing this work forward.

Conclusion

This chapter explored the ways young people transform from environmental or reproductive health activists into population-environment advocates. Along the way, this transformation opens up new possibilities for them to assert a unique role as leaders of a global movement, and allows them to speak with authority on global development problems and solutions. At the same time, key moments emerged when it became clear that advocates were ambivalent or uncertain about the discourses underpinning population-environment linkages and family planning solutions. This uncertainty often heightened even further when discussing the issues of race and reproductive justice. The next chapter explores how populationists deploy the language of reproductive justice, through an investigation of the striking absence of race in their discourse. Specifically, it engages questions about the absence of analysis of race and class issues in population advocacy, and how this erasure is central to populationists' self-defined social justice agenda.

5

Co-Opting Reproductive Justice

The reproductive justice legacy in the field is a strong positive one for most people, and therefore it's a good train to get on.
—Pamela, private foundation donor

Anything done about us, without us, is not for us.
—Loretta Ross, former National Coordinator, Sistersong Reproductive Justice Collective

Reproductive Justice (RJ) is an American social movement led by feminist women of color activists. In 1994, having returned from the ICPD in Cairo where they organized and collaborated with Global South feminists, reproductive rights activists of color began to outline an organizing structure based on applying the human rights framework invoked in Cairo to American reproductive politics.[1] At the heart of their politics was a response to, and rejection of, the long history of reproductive oppression in the U.S. based on race, gender, and poverty. Racist population control, asserted in the form of coerced sterilization, surveillance, and governance of low-income women on welfare, and discourses and narratives that rendered women of color's fertility, childbearing, and motherhood problematic, even criminal, have been central components of reproductive politics in this country. Within this history, it is abundantly clear that the discourses of private, individual choice and decision-making that dominate the narrative of mainstream reproductive rights are far more complex than they appear in that they are not, nor have they ever been, available to all women. As a result, the founders of the RJ movement de-center individual, private choice, abortion, and contraceptive access in their activism, replacing them with a comprehensive focus on the social, political, economic, and cultural contexts that shape women's reproductive lives.[2] Within this comprehensive focus, RJ activists frame the movement around the centrality of the right to have children as well as the right not to have them, and to raise them with the necessary social resources to do so. They also declare an anti-Malthusian position, given its long history of justifying coercive human rights abuses against women of color around the world.[3]

This chapter foregrounds these foundations in order to contrast the central tenets and commitments of the RJ movement with how the framework has been taken up by the Sierra Club and their partners through GPEP trainings. GPEP staff and volunteers frequently presented themselves as RJ activists. They described their RJ commitments through language supporting women's rights to bodily autonomy and freedom from coercion, even as they espoused the contradictory populationist view that the fertility and childbearing of the poor is a driver of environmental problems. In my interviews of GPEP staff and observations of their trainings, what clearly emerged was that the Sierra Club's use of RJ co-opts the language of the framework, while sidestepping the uncomfortable racial politics at the center of the social movement. However, this is not the focus of this chapter. The chapter is concerned with the work that this co-optation accomplished in the minds of GPEP leaders and partners. My argument here is that using RJ operates as shorthand signifying progressive politics among populationists in ways that help them navigate thorny racial politics without addressing them directly. In other words, the Sierra Club and its partners use RJ as what Cornwall and Brock refer to as a development buzzword.

Cornwall and Brock trace an increasing trend in international development toward using seductive "buzzwords" that "promise an entirely different way of doing business" and that "speak to an agenda for transformation that combines no-nonsense pragmatism with almost unimpeachable moral authority."[4] Key words like poverty reduction, participation, and empowerment are seductive precisely because they invoke a sense of optimism and an alignment of development mission, policy, and action that creates positive change. They also invoke a moral goodness central to "conferring on their users that goodness and rightness that development agencies crave and assert in order to assume the legitimacy to intervene in the lives of others."[5] Reproductive justice has become one of these key terms—it is used frequently in population advocacy materials and in advocacy trainings. The myth about development buzzwords is that they can transcend context and politics and have a sense of universal meaning—which is precisely why they operate successfully. This chapter exposes this myth in the context of populationist framings of RJ, analyzing the ways proponents use the framework to represent their work as progressive, even as they express ambivalent perspectives on RJ behind the scenes.

The chapter is organized as follows. First, I explore how RJ language has been deployed within the GPEP, as compared with RJ organizations, and the contradictions and confusions this produces. Following that, I explore the history of reproductive politics giving rise to the RJ movement, and the tensions and fault lines that opened up between mainstream, white-led women's reproductive health organizations and those led by women of color. Last, I analyze some of the ambivalence population advocates express about RJ behind the scenes, even as they invoke its language in attempts to build coalitions and to secure a progressive framework for their efforts.

Making Population-RJ Activists?

Nora, an African American woman in her mid-20s, had been volunteering as a GPEP training facilitator for several years. She was one of very few African Americans involved with this kind of advocacy, which she felt acutely when talking to women of color activists, as well as when dealing with other members of the population-environment network. At a certain point, her sense of isolation and conflicted thoughts about race, RJ, and population activism led her to consider leaving the work altogether:

> Last year I was having my own existential crisis about organizing around population issues and I got to the point where I was like, I don't want to talk about this. I want to talk about other things. It's been hard for me. Like if anyone was to ask me, then I'd say that I'm a reproductive justice activist first. My struggle last year was, I was getting more involved with RJ activism, and kind of learning more about people of color activism, and was recognizing that it has this horrible history, that population, like what people want to do about it and how they want to talk about it. So it was an existential crisis for me. I think a lot of people think about that

> horrible stuff from the past and they're like, "well we're just not going to talk about it at all," that's like when people feel that because the US has a horrible history with racism, they often don't want to talk about it, but I feel that's all the more reason to talk about it. So I'm still on the side of, "let's have a conversation about population that's a culturally competent and intelligent conversation."

Nora considered herself to be a sexual and reproductive health activist first, and she often found that when she went to women's health events and spoke about population issues, others in the room made assumptions about her, or wouldn't talk to her at all. The assumption that she was a neo-Malthusian sympathizer particularly stung. A year or so prior, Nora had attended an annual young women's reproductive justice gathering at Hampshire College known as the Civil Liberties and Public Policy, or CLPP, conference. CLPP has a strong radical feminist vibe; unlike Sierra Club events, it is very racially diverse, and race and class critiques of reproductive politics are high on the agenda. While attending CLPP, Nora was brought face-to-face with the political stakes of what she was doing:

> I went to the most recent CLPP conference at Hampshire College, and to be honest that's when I started to think that, kind of, I felt that existential crisis. Because I went there and I saw the reception that people got from talking about the Sierra Club and talking about reproductive rights. There was a woman there with an organization that talks about human rights and about population, and I was just really wanting to talk to her and thank her for being there, because the hostility coming from some of the people on the panel with her, you know, can you really have a conversation about population? Of course they were cordial and polite, but they were not really into what she was saying, so, that really put me off. And so it was a tense situation and it wasn't really positive.

Her sense of racial isolation and confusion raised deep questions for Nora over where she stood with respect to population, reproductive health, and RJ. Could she have conversations about global population without aligning herself with population controllers? Would she always be isolated from other black women in RJ spaces? Given that mainstream environmental organizations like the Sierra Club are often dominated by white leadership and membership, her sense of racial isolation was complicated by a sense of hyper-visibility when she attempted to address racial issues with her fellow population advocates: "The conflict over history and the bad things that happened is not over, and people are still having conversations that don't acknowledge race. So it's still a struggle, and it's something that I don't feel like I should have to bring to the table, but I do think that people kind of expect me to be the one."

Briana, another young African American population advocate, echoed Nora's sense of racial isolation in environmentalist spaces, which only increased when population issues came up:

> At the [One Voice] summit, they asked, "How many people are here to address these issues through environment?" I was the only Black girl there raising my hand. Most other Black girls there didn't raise their hands, and when I asked about it they said they'd never thought about it . . . Learning environmental studies is depressing sometimes. Depending on who talks about environmental issues, like if it's an old white guy, I'm skeptical. But if they're talking about sexual health, etcetera, then I'm like yeah. When I've gone to Sierra Club events, they're all white. I'm always the only person of color.

Nora's and Briana's comments raise a question that plagued me throughout the project: the question of positionality. The stakes of population politics are high; as many advocates told me over and over, population advocates often feel that they are being attacked by people on the political left and the right. Leftists make charges of racism and neo-Malthusianism; those on the right reject their stance on universal access to contraceptives and abortion. I also found, over the course of this project, that any proximity to population work—even in the context of research—leads to assumptions and questions about one's position on the issue. On many occasions, population advocates assumed that my goals and priorities aligned with theirs simply because I was in the room. When I spoke to fierce critics of populationism, they assumed that most populationist advocates actually support population control, while advocating women's SRHR. Of course the truth is a bit more complicated than that. The majority of populationists I interviewed identified as feminists; they saw family planning as a necessary tool to give women the freedom to make a range of choices in their lives, predicated on being freed from the burden of unwanted childbearing. At the same time, they expressed these views in a neoliberal framework, whereby reproductive decisions are privately made at the individual level, chosen freely, and based on access to contraceptive technology. While they acknowledged the constraints of poverty, they saw this constraint as primarily preventing access to tools and services, rather than structuring the conditions of women's reproductive lives.

Many of the youth advocates I spoke to were uncertain about what RJ truly meant; they were also confused or uncomfortable with the racial history of population control, and what that said about their advocacy work. However, its positive connotations were able to smooth over some of these concerns. One young South Asian woman stated that while she was uncertain what RJ actually meant, she was certain that it was a good thing:

> I think fundamentally, everybody should have a right to . . . access to contraceptives, and everyone should have the right to have as many kids as they want, and control over their fertility. Reproductive health justice or reproductive justice, I don't know what it is exactly, it has, I think, a lot more popular, uh, more people know about it and understand what it means. To be talking about reproductive health. I think at least in my world health disparities is so well known and well understood. When you say something like reproductive health justice I think it automatically makes people understand . . . reproductive justice, in terms of like reproductive health care for everyone.

Other youth advocates expressed similar confusion during interviews, despite using the terms "reproductive justice" and "reproductive health" frequently and interchangeably at advocacy events. This is perhaps because in GPEP trainings, participants were expected to openly take a stand on the issues through interactive activities and discussions. Through role playing and values clarification exercises, marking out one's position was a requirement of being in the room. It was, at times, an uncomfortable place to be, but also a necessary component for thinking through how to craft a clear and specific message later. Over and over, during these exercises and the conversations that arose afterward, the type of discomfort Nora and Briana expressed came to the fore.

Taking a Stand

On a hazy June morning in San Francisco, another advocacy training workshop was getting under way at the Sierra Club headquarters. This time the goal was to bring together activists across a range of age groups, based on a GPEP leader's desire to engage adult activists who were already involved with the Sierra Club, or who had a previous interest in either environmental or reproductive health and justice issues. As she mentioned in an e-mail, "One thing that our program is working towards in the coming year is increasing our visibility amongst the environmental and specifically Sierra Club community, and this training will help us move towards that goal."

Entering the conference room, I took note of the crowd: approximately thirty people, predominantly white and female, seven or eight visibly people of color. Half the crowd appeared to be in their twenties; of the rest, many were over fifty. The day's session was co-presented by a facilitator from a national women's reproductive health organization. Facilitators laid out the goals of the training, as they had in the many GPEP workshops I had attended before. This time, however, population control was mentioned briefly as a component of RJ debates, an element of the "dark past" that must be addressed in order to make sure that today everyone can exercise their full capacity to determine when, where, and how to reproduce. Immediately following, though, the facilitator moved on, saying, "I don't want to spend too much time on that because we have a lot to do today." She then launched into a description of the reason we were there: to learn about international family planning. She introduced family planning as a comprehensive set of services, not just contraception and abortion, but also maternal and infant health care, full reproductive and sexual health, and STD and HIV/AIDS prevention. The main issue, we were told, is access: choosing whether and how to reproduce or even seek services, tools, and technologies. Individual access, more than anything else, was the crux of the issue.

From there, we moved directly into a values exercise, in which a facilitator read a list of statements and instructed participants to line up in various places in the room to indicate agreement, disagreement, or a middle position—"in between"—in response to the statements. At the first statement, "U.S. actions cause the most global ecological degradation, thus people in the Global North should not address population issues," most participants moved to indicate their disagreement. On other statements, such as "All environmentalists should work to address issues of reproductive health and rights" and "All reproductive health and rights activists should work to address environmental issues," the majority of participants moved to indicate that they agreed.

Despite the earlier facilitator's description, in these statements, reproductive health and rights were framed to indicate access to contraceptives alone. Broader issues or services related to maternal and child health, prevention and treatment of sexually transmitted infections (STIs), prenatal and postnatal care, and fertility services were curiously left out—and there was very little commentary. When the facilitator asked whether there were questions or reactions after each values statement was read, participants were generally quiet. The mood in the room was polite and muted. During the exercise, several participants held side conversations with those who moved to the same side of the room as them, but these whispered asides were not shared with the larger group.

This shifted slightly when a participant broached the topic of race. The facilitator read a statement: "Everyone on Earth has the right to live at the same standard as how people live in the U.S." Most people moved to the "in between" position; when asked, several people told the facilitator that even people in the U.S. shouldn't live at U.S. standards. In the midst of these comments, one of the few African American women in the room stated that she didn't know what a U.S. standard of living is, because many poor African Americans live at the standard of people in developing countries. This was met with chilly silence. No one responded; the awkwardness of the silence was palpable. We moved through the remaining values statements, ended the exercise, and returned to the table.

In a side conversation later in the afternoon, a participant mentioned to me that she would not return the next day. "I'm not into this stuff, telling poor women they shouldn't have children." She was in her fifties, white, and a longtime reproductive health activist. I asked what prompted her to attend the training. "I thought I'd learn something new.

And I'm interested in the environment, and new ways to connect it to reproductive health. But this stuff? This isn't new. This stuff is really old. We used to hear this stuff all the time in the seventies. I can't believe they're still trotting these ideas out today." I thought about her comments as I returned to the next session, where the presenter focused on contextualizing women's reproductive health and rights as barometers for measuring progress on social issues—particularly in a globally interdependent society. Like the earlier presenter, she referenced population control briefly, but brushed it aside, citing time constraints. For the rest of the afternoon, she led a session emphasizing "basic rights" as universally important and beneficial, arguing that achieving these rights would move us forward on reproductive health, environmental sustainability, poverty alleviation, and economic development in the Global South. Focusing firmly on universal language of shared goals and benefits, she told us to keep the notion of global interdependence in the forefront of our agendas, along with a desire to keep in account all issues that impact others. Questions of justice were tabled for the day, and for the rest of the training, they did not reappear.

Foregrounding Race and Rights

The ways race was—and was not—addressed in the session were particularly striking when compared to an RJ conference I had attended several months earlier in Washington, DC. It was an annual membership meeting and advocacy event organized by the Sistersong Collective for Reproductive Justice. Emerging from the Union Station train stop and joining several small clusters of African American and Latina women, I rushed toward the hotel ballroom, conference tote bag, folder, and handout materials in tow. We entered the space to find over four hundred people, mainly women, and although I had attended multiple reproductive health advocacy events at that point, this was the first comprised predominantly of women of color. The energy in the room was warm, vibrant, and informal. Participants received an ebullient welcome from the stage, to which they responded with frequent bursts of applause and cheers. Looking around, I noticed women of all ages, from late teens to mid-seventies. I also noted that not one member of the population-environment network that I had spent so much time with over the prior year was in attendance, despite repeatedly expressing their commitment to RJ.

Despite the festive vibe, I had begun to feel some level of conference fatigue. This was the fifth SRH-related advocacy event I'd attended in a six-month period, and after a while some of the messages and vocabulary began to run together. Contraceptive access, abortion rights, women's autonomy and empowerment, justice, the centrality of policy, and the importance of congressional lobbying are all emphasized in Sierra Club trainings and other population-environment advocacy events, just as they were by Sistersong. Hearing this language over and over, witnessing the whipping up of activist fervor around recent small-scale policy victories and the moans of despair at congressional votes on women's health services legislation, after a certain point began to feel scripted. After all, regardless of thematic focus, conferences organized around women's SRH share a common parlance, and a common sense of urgency: women's rights are under threat, must be expanded and protected, and we, the women in the room, must work at the forefront of protecting our and other women's access to high-quality reproductive health services.

However, this was where the similarity ended. Unlike the population-environment meetings I'd attended, the Sistersong conference emphasized articulating RJ concepts through intersectionality[6] and resistance to racialized reproductive oppression. Race, class, and gender were at the heart of Sistersong's analyses, along with critiques of intra-movement dynamics within SRHR advocacy, particularly the differences in approach between mainstream organizations and those led by women of color. During a lunchtime keynote speech, the speaker articulated the ways reproductive justice (movement building), reproductive rights (legal advocacy), and reproductive health (service delivery) were inextricably linked. As the speaker emphasized again and again, RJ is predicated on the interrelationship of all three approaches, including the direct rejection of all attempts to erode the rights of women, particularly women of color and the poor. In a powerful moment, she shared a personal experience of coerced sterilization, arguing that population control is alive and well in the lives of women of color and low-income women today. In a context in which the ability to give birth and parent children is not assured, she argued, contraceptive access and abortion rights are not the primary issues for RJ activists—instead, the center of the movement is the ability to make autonomous decisions based on a foundation of human rights.

In a conversation later that week, it was clear that a GPEP manager at the Sierra Club was confused about the distinctions between reproductive health and RJ, and used the terms interchangeably:

> Some people don't want to take a more nuanced approach, but with them, we emphasize effectiveness. We say that the reason to be passionate about a cause is because you want something to change. We tell them that feeling passion for this issue, making a difference, gaining traction with environmentalists, policymakers, reproductive health activists, the reproductive justice community, you have to think about things in a nuanced way in terms of making change in the world . . . I interact daily with environmental justice and reproductive justice groups like Planned Parenthood and Choice USA. Partnerships are really important. We wanted to bring broader perspectives through this training and through campus tours. We always partner in our activities with organizations that do RJ work.

I mentioned this confusion to a longtime feminist critic of reproductive politics, who was unsurprised at the jumbled use of terms, noting that the realities of funding constraints in women's reproductive health often lead organizations to take what seems to be the most expedient path. She stated further:

> There's a somewhat naïve belief that you can have your cake and eat it too, that you can be pro-reproductive justice and anti-population growth. Reproductive justice is getting watered down and substituted for reproductive rights, it's a watering down of language. I don't believe that they're cynical, I think the more radical meaning of reproductive justice calls into question the linking of reproductive justice and population. The purpose of some of these programs is to increase funding for international family planning but sometimes I don't think they believe what they're saying. It's just strategic. The rhetoric ebbs and flows according to the fads of the times.

How could such a radical departure in language and framing be accomplished, particularly given the RJ movement's critical anti-populationist roots? Is this just a cynical use of development buzzwords?

Buzzwords may be symbolic, but their use has material effects; they are closely linked to policy-shaping strategies. This is because they frame problems in ways that define particular paths of action, or intervention. As a result, buzzwords have increasingly become necessary components of how development projects define and shape their work: "Nobody trying to be influential can afford to neglect the fine art of buzzwords . . . Images conveyed by simple terms are taken as reality, and words are increasingly loaded with ideological symbolism and political correctness. It may seem innocuous. It surely is not. Why make a fuss? The reason is that the terms we use help to shape the policy agenda . . . The linguistic crisis is real, and is not going to go away."[7]

In a field like population and family planning, which has had a controversial history, RJ operates nicely as a buzzword, conveying a positive connotation and keeping population discussions from being mired in the ugliness of racism and coercion. It also helps that advocates relegate population control to the past—the "dark past," in the parlance of advocacy trainings. And yet, these historical foundations are rarely, if ever, explored in trainings. Racial politics in the U.S. never entered the discussion during GPEP activities, with the exception of the moment when a training participant raised them, whether because they would be seen as irrelevant to international policy, or because the Sierra Club itself has its own history of racial controversy.[8] However, an understanding of the context from which RJ arose is necessary to understanding how it operates alongside mainstream reproductive health and rights activism today, and why the differences in approaches are deeper than semantics and policy focus. They are, in fact, centered on the differential histories of reproductive politics in the U.S., and how those politics have been heavily shaped by race and class.

Reproductive Justice in Historical Context

In the U.S., family planning has played a prominent role in the mainstream political debates over women's sexual and reproductive health in that contraceptive research and access, and secure legal rights to abortion, have been central issues at the heart of feminist organizing. At the same time that hard-won reproductive rights were being achieved for many white women in the U.S., those same rights were being eroded for American women of color, and those in Global South countries, under the same banner of birth control.[9] For middle-class, white American women, birth control access was expanded in the name of opening up increased social freedoms for women. For poor women, women of color, and those in the Global South, this access was often expanded and coercively imposed under an agenda of population control.

Rather than occupying a shared ground for women's organizing across social groups, the topic of birth control has served as a site for social and political fracturing within the women's movement in the U.S., primarily along race and class lines. Advocacy for "voluntary motherhood" in the late 1800s radically challenged what had been seen as men's right to have their sexual and reproductive urges fulfilled by their wives at all times.[10] A century later, calls for access to safe, effective contraceptive methods and legal abortions were no less politically contentious; however, they had gained broader ground within feminist organizing, such that access to birth control, including abortion, had become a central issue in mainstream women's rights organizing. At the same time, the focus on birth control exposed deep tensions and hostilities within the women's movement, exposing the racial, religious, and class-based fault lines around which birth control debates have historically been organized.

Issues of sex, pregnancy, childbearing, and motherhood—what Solinger refers to as reproductive capacity—have always carried different meanings for women and girls in the U.S., based on race, class, and historical moment.[11] The history of reproductive politics in this country is one in which policies, laws, and other programs designed to intervene on reproductive capacity have been shaped by concerns about how to solve broader social, political, and economic problems facing the country—whether the problem was to increase the labor force, as in under slavery, or to decrease the number of people accessing public financial resources (i.e. welfare). Women's reproductive capacity has historically been viewed as a national resource that could strengthen the nation or weaken it. Black women's history in particular has been marked by the "systematic, institutionalized denial of reproductive freedom," from slavery to the present.[12] The very notion of reproductive freedom in the U.S. is constituted by notions of race, along with the individualized notion of reproductive choice.

Roberts notes several examples of how black women's fertility and childbearing have been criminalized in the U.S. In 1989, police began arresting drug-addicted pregnant women, targeting and arresting those in poor neighborhoods, as well as staking out maternity wards and taking women to jail immediately after giving birth. During the operation, close to fifty women were arrested; all but one were black. The following year, an editorial in the *Philadelphia Inquirer*, "Poverty and Norplant—Can Contraception Reduce the Underclass?," argued that the primary reason for black childhood poverty is that "the people having the most children are the ones least capable of supporting them," and recommended coercively implanting women with long-acting reversible contraceptives to prevent pregnancy.[13]

This is not to suggest that birth control is not an important component of women's bodily and reproductive autonomy. However, leaders of the birth control movement in the U.S. have historically overlooked the needs of lower income women, many of them women of color, as well as playing into racist constructs. As Davis so cogently explains, in the 1960s and '70s, women of color were frequently absent from mainstream abortion rights organizing efforts, not because they were unaware of or uncommitted to the issues, but because of classist ideologies embedded within the birth control movement itself. For example, the movement often blurred the distinction between the right to legally access a safe abortion and the abortion itself. Abortion occurs within a matrix of social and economic conditions that facilitate or mitigate against the viability of bearing and raising children. These conditions are often the conditions of poverty, and having abortions alone does not lead to having a decent job with steady income, quality health care, safe housing, good schools, etc. Focusing on abortion alone obscures the context within which it has been necessary for so many women—and without which they may have chosen to have more children.[14]

While many white feminists attributed women of color's resistance to participation to a lack of consciousness, or lack of desire to engage in social movement organizing across racial lines, in actuality resistance was rooted in the fact that women of color faced a very different landscape of access and control. Where middle-class white women were struggling for the right to limit their fertility, poor women and women of color were fighting forced sterilization and other imposed means of coercive fertility regulation.[15] In other words, for women of color in the U.S., the struggle for birth control rights has been a struggle to maintain reproductive autonomy, both in the context of the right to bear children as well as the right to avoid bearing them. A narrow focus on controlling and limiting one's own fertility, long fought for as a "right" for middle-class white women in the U.S., has historically been imposed as a duty on women of color.

After abortion was legalized by the Supreme Court via *Roe v. Wade*, an immediate backlash followed, giving rise to the Hyde Amendment in 1977, which mandated withdrawing federal funding from abortion services. Most states followed suit, causing many low-income women to lose their access to abortion services. Meanwhile, surgical sterilizations were free on demand and funded by the Department of Health, Education, and Welfare, leading many low-income women of color to choose sterilization. At the same time, thousands of low-income women of color were sterilized without their consent. Shocking news reports began to surface, detailing the fact that doctors and medical students across the country were performing coercive surgical sterilizations on women of color, many of them on welfare. In one particularly egregious example, in June 1973, fourteen-year-old Mary Alice and twelve-year-old Minnie Lee Relf underwent tubal ligation surgeries in Montgomery, Alabama. Their mother, unable to read and signing her name with an X, had consented for her daughters to have vaccinations, which she was told the girls needed for their health. However, the agency that had recommended the vaccinations also operated a family planning clinic, and it decided to go a step further by sterilizing the girls. Based on Mrs. Relf's written "consent," the agency performed tubal ligation surgery on Mary Alice and Minnie Lee. After the Relf case surfaced, and following a public outcry, news came to light that federally sponsored birth control clinics had also sterilized at least eighty other minors in a fifteen-month period, mainly in southern states. The Relfs, like many of the other minors' families, were on welfare.[16]

Between 1964 and 1973, there were documented cases of more than one thousand poor women, most of them African American, who were involuntarily sterilized in the U.S.[17] Thousands more sterilizations were carried out across the country throughout the twentieth century under eugenic sterilization laws, with most of those sterilized marked as either mentally ill or mentally deficient.[18] Yet, the sterilizations of poor women of color were different from other eugenic sterilizations in that race, poverty, and being perceived as a drain on state resources were used as primary reasons to justify the procedures.

Sterilization abuses of Mexican and Chicana women in California were revealed in an exposé article published by Claudia Dreifus in 1975.[19] Dreifus, along with Dr. Bernard Rosenfeld, undertook a series of interviews with doctors at USC-L.A. County Medical Center, where coerced sterilizations had occurred. Dreifus did not tell any of the doctors that she was a reporter or that their words would appear in print—an ironic fact, given that she and Dr. Rosenfeld were investigating abuses of informed consent. Nevertheless, what they found was shocking; their major conclusion was that "forced sterilization is a part of academic training at more than a few major teaching hospitals around the nation."[20] Dreifus noted that the doctors they interviewed "seem[ed] to accept coercion as an everyday fact of medical life—few of them [were] even aware of the moral significance of what they have witnessed."[21] Of the twenty-three doctors they spoke to, nearly half had witnessed coercion, or related conditions: "hard-selling, dispensing of misinformation, approaching women during labor, offering sterilization at a time of stress, on-the-job racism." Only four doctors were able to definitively say that these practices did not occur at the hospitals where they trained. Doctors told stories of sterilizing black teenagers in inner cities as young as sixteen years old, patients they described as having a "low mentality," whom they would offer tubal ligations to while they were in labor. If women had three or more children and were poor, doctors would "sell the operation." One doctor described how interns were often motivated by simply wanting to gain more experience performing the procedure—"That was a big influence in prompting them to do it—they wanted to get another tubal under their belt."[22] The doctors revealed that they were more likely to push the procedure to women with multiple children who were on welfare. In some interviews, doctors who had done their training in the South (North and South Carolina and Mississippi) revealed that teenage daughters of welfare recipient mothers were routinely sterilized.

As these examples demonstrate, the notions of individual choice and private reproductive decision-making ignore the fact that economic constraints and politics have placed significant constraints on reproductive "choices." The notion of choice in the mainstream reproductive health movement assumes the kinds of social and racial privileges that offer freedom from constraining state policies and interventions. Women make reproductive decisions within the context of specific social, economic, political, and cultural contexts. These conditions shape and constrain the ways women envision what is possible, access resources, and are excluded from them. At the same time, the notion of choice assumes no role for the state, which history demonstrates has played a major role in regulating the reproductive lives of women of color in the U.S.

Also in the 1970s, the Indian Health Service sterilized at least 25% of Native American women between the ages of 15 and 44, in order to close the gap in birth rates between Native communities and the national median. Women were not provided full access to information about sterilization, did not give proper informed consent, and were not given a 72-hour waiting period between giving consent and having the procedure performed. Doctors pushed the procedures, threatening women with the removal of their children and their welfare benefits.[23]

In addition, Native American women have historically been plagued by the possibility of forced removal of children from their homes. Beginning in the seventeenth century, white colonizers began creating Christian schools that would separate Native people from their communities to give them a "civilizing" education.[24] In the late nineteenth century, however, these schools were formalized as part of federal policy overseeing the administration of reservations. The policy mandated that Native children be taken far from their homes at an early age and educated in Christian schools; they weren't returned to their families and communities until early adulthood. By the late twentieth century, more than 100,000 Native children and teens were educated in these schools, both on and off the reservations, where the central ethos was to "kill the Indian and save the man."[25]

When women of color activists sounded the alarm about these injustices, they found that many of the outspoken white women who were reproductive health activists refused to stand with them, for example refusing to advocate against sterilization abuse. While women of color were coercively sterilized, white women had a hard time accessing *voluntary* sterilization. After hospitals established restrictive sterilization rules in the 1950s, in keeping with postwar pronatalist sentiment, they adopted restrictive policies such as the 120 rule, which stated that those whose age and number of children multiplied together must reach at least 120 in order to qualify for surgical sterilization.[26] Many hospitals adopted a 150 or 175 rule to protect themselves from lawsuits. These policies disproportionately impacted white women—the group who was targeted for postwar pronatalist efforts. At the same time, in the 1950s, southern physicians began to perform unlawful forced sterilizations on healthy women of color. They also chose not to adopt the same age/parity policies developed in other regions so that they could continue to forcibly sterilize low-income black women. As a result, "only white women outside the South complained that they had been denied voluntary sterilization."[27] White women's complaints about lack of access to voluntary sterilization led the ACLU, ZPG, and the Association for Voluntary Sterilization (AVS—now known as EngenderHealth) to launch a campaign through lawsuits in 1971 to overturn hospital restrictions. Plaintiffs in all of the lawsuits were white women.

In order to respond to abuses and to assert agency and leadership, women of color activists organized formally, on a national scale. In 1992, the Women of Color Coalition for Reproductive Health Rights (WOCCRHR) was formed, consisting of six organizations, including Asians and Pacific Islanders for Choice, National Black Women's Health Project, National Latina Health Organization, Latina Roundtable on Health and Reproductive Rights, National Coalition of 100 Black Women, and Native American Women's Health and Education Resource Center. This coalition organized to play a role in the 1994 ICPD, or Cairo Conference. Members of the group who participated in the ICPD delegation returned to create the U.S. Women of Color Delegation Project and issued a statement drawing linkages between their experiences of reproductive oppression and those of women in the Global South. The group connected their struggles to transnational struggles for poverty alleviation and women's rights, and asserted that women in the U.S. experienced similar human rights abuses and population control as women in developing countries. In returning from Cairo, group members shifted their focus into incorporating a human rights framework into their efforts, and created a coalition named the Sistersong Collective for Reproductive Justice.[28]

They articulated RJ as a way to bring the human rights framework home from Cairo and apply it to the reproductive abuses experienced in the U.S. Like other activists at Cairo, they had advocated for women's human rights, voluntary access of contraceptives, and comprehensive reproductive health care, and a rejection of the population control paradigm in international development. However, they also turned the lens on the U.S., identifying coerced sterilizations of women of color as human rights abuses and connecting them to the experiences of women abroad. Since that time, two central foci have framed the RJ movement: asserting the leadership and self-determination of women of color; and reinforcing the inextricable links between the importance of the right to have children as well as to not have them, and to parent existing children with the necessary resources and conditions for doing so. Central to this approach is a grounding in human rights, predicated on the reality that reproductive rights can only be fulfilled in a context of civil and political rights (including legal recognition in courts and freedom from forced pregnancy, domestic violence, and repressive religious codes) as well as economic and social rights (such as accessible health services, safe nutrition, and contraception).[29]

Shortly after women of color activists developed the RJ framework, mainstream women's organizations began to draw on it, occasionally integrating RJ language into their own work. For example, the 2004 March for Women's Lives, initially organized by Planned Parenthood, Feminist Majority Foundation, NARAL Pro-Choice America (NARAL), and the National Organization for Women (NOW), became a coalition-led event when Sistersong was asked to endorse the event. In negotiating their participation, Sistersong leaders insisted on steering committee participation for women of color (none of the mainstream organizations had women of color in senior leadership, thus there were no women of color on the steering committee), a broadened framework emphasizing the full context of women's reproductive lives, rather than a narrow focus on abortion rights, and a new name for the march that reflected this perspective.[30] The coalition accepted the shift, and announced it publicly for the first time in an e-mail from NOW that emphasized the importance of "demanding political and social justice for women and girls regardless of their race, economic, religious, ethnic or cultural circumstances," in an effort to "move forward with full equality and reproductive justice for all."[31]

Following this event, it was unclear whether mainstream women's organizations would continue to support or fully integrate RJ frameworks. While certain organizations continued to use RJ language, the central tenets of human rights advocacy and a community-based contextualization of women's reproductive lives quickly dropped away from partner organizations' websites and materials. More than a decade later, RJ language continues to be used by mainstream women's organizations, in part to attract younger and more diverse supporters, but the linked role of intersectionality and human rights is often conspicuously absent.[32]

Reframing RJ?

As mentioned in chapter 4, the GPEP approach to training youth advocates was based on a justice framework gleaned from a book called *A Pivotal Moment: Population, Justice, and the Environmental Challenge*. The book claims to draw from RJ to frame the concepts of inequality and justice that drive population growth; however, in an interview with one of the framework's authors, when I asked how EJ and RJ groups have responded to the notion of population justice, she made it clear that she did not have direct ties to those communities: "Good question. I haven't done any outreach to those groups. We had some EJ people at our group, like one woman in particular who is also working on climate justice, and that's certainly a part of the book. One of the authors approaches climate change from the perspective of, what if everyone on earth had equal access to the atmosphere? But that's a good question and I don't know the answer."

However, she did not see this as a hindrance from engaging with the language and politics of RJ. During a panel discussion at a population-environment workshop, she drilled down on the connections between population and justice by acknowledging race and class, but failed to make the connection between RJ and critiques of populationist perspectives like her own:

> The best way to slow population growth is by ensuring the means and power for people to control their own childbearing. The reproductive

> justice movement says that we must look at the totality of people's lives, including race, gender, and economics. Inequalities that limit decision-making. Poverty has a huge impact on decision-making. Reproductive justice means that people have real power to make their own decisions.

Many of the people I encountered in the population-environment network had limited knowledge of RJ, or were more hesitant to integrate the framework into their efforts. As previously mentioned, youth activists in particular expressed uncertainty about what the term meant, although they viewed it as positive. However, Marisa, the former GPEP manager, expressed a more ambivalent position. She frequently invoked RJ in her presentations and advocacy trainings. Yet, behind the scenes she expressed deep frustration with the framework, arguing that a strict adherence to RJ principles is actually an impediment to population-environment work. For example, she referred to Nora, the young woman whose experience opened this chapter, expressing annoyance at how feminists at the CLPP conference were unreceptive to her message: "She attended the Amherst group's [CLPP] conference . . . she was blown away by how much they attacked the Sierra Club for addressing these issues. In one of the conferences, they said you should never get involved with that population program, referring to us. And our volunteer reported all this back to me, and she said she felt really uncomfortable and it made her question her values."

I asked whether she thought that the CLPP conference participants were focused on addressing histories of reproductive coercion in the U.S. She thought so, and argued that this framing is limited and unhelpful because it focuses too much on dispelling population myths: "They're so focused on the past but they're not interested in looking at solutions now. When I talked with them a few months ago, it was like, well what solutions do you support? I think that they organizationally support reproductive justice solutions but from the critical population issue they won't work with us. Paul Ehrlich's message did a lot of damage, but that's in the past, and where we're moving with this campaign is to change the dialogue, to change the messaging, to move forward in a more progressive way."

However, population control is not a thing of the past. A 2013 report revealed that doctors working for the California Department of Corrections and Rehabilitation sterilized 148 female prisoners via tubal ligations between 2006 and 2010 without obtaining the required state approval for the procedures. There were perhaps 100 more similar procedures performed, dating to the 1990s.[33] Doctors appeared to have targeted those who had been to prison multiple times; one doctor in particular connected the procedures to state welfare. While denying that he had coerced women into obtaining the procedure, he argued that the compensation doctors received for the surgeries was minimal compared to the savings for the state: "Over a 10-year period, that isn't a huge amount of money compared to what you save in welfare paying for these unwanted children—as they procreated more."[34] Ironically, another physician described the procedure as "an empowerment issue" in that offering tubal ligations was a way of "providing them the same options as women on the outside."[35]

Despite the deep tensions and conflicts involved in the struggle over RJ, its languages and frameworks are continually taken up by groups whose work does not address race or class. Given the easy adoption of RJ as a development buzzword, it can easily be used to neatly align with populationist arguments. As this chapter demonstrates, the meanings and uses of "reproductive justice," "social justice," and "human rights" are constantly being reworked, negotiated, and at times transformed in unrecognizable ways by competing actors and narratives.

Conclusion

This chapter has revealed the complicated ways RJ language is deployed in the contrasting advocacy work of populationist organizations and RJ organizations. Populationist organizations drew on this language for strategic reasons, to try to advance their agenda while not being hindered by racial controversies. And yet, as the history of RJ organizing demonstrates, critiquing racism in reproductive politics is central to why the movement was founded. When populationists utilize this language, they not only accomplish a linguistic sleight of hand, they also co-opt a long legacy of social activist work done by women of color who have rejected precisely the kinds of strategies groups like GPEP adopt. Yet, in the outcomes-based world of policy advocacy, this is perhaps irrelevant. RJ, like women's empowerment, has in many ways been offered as a simplified way for populationists to position themselves as doing progressive work.

Conclusion

Is There a Feminist Way Forward?

Several years ago, multiple online news outlets published an article written by an American woman who described herself as a GINK: Green Inclinations, No Kids.[1] She argued that the global population problem is a problem of affluence-based resource use, driven by people like herself: white, American, and middle class, and she reasoned that she, and others like her, should consider not having children if they wanted to make a dent in environmental problems such as climate change. At the time, I thought that the article offered a fresh perspective in placing global environmental problems on the shoulders of race- and class-privileged people, while explicitly decentering narratives blaming people of color, immigrants, refugees, and populations in the Global South. The author was clear that much of the population discourse that circulates in the environmental realm is about blame, and that the blame has been misplaced; she instead adopted a stance of embodied responsibility for the problem, saying "steer that blame right over here." Yet she was still making a populationist argument. And in doing so, even as she argued against blaming environmental problems on the reproduction of the poor and marginalized, she still located the source of the problem in women's wombs—a framing that is simplistic and inaccurate. Perhaps we should steer that blame away from women's wombs—*any* women's wombs.

These are difficult questions to grapple with, particularly for feminists who care about the environment and about women's sexual and reproductive health and autonomy. It is difficult to resist slipping into populationist logic—how can we not deal with the population question? Is it enough for feminists to insist on women's bodily autonomy, or must we also find a feminist way to grapple with population growth as an environmental problem? Can feminists be populationists too? There is no singular answer to this question, but feminists are grappling with it in increasingly complex ways. In *Staying With the Trouble: Making Kin in the Chthulucene*, Donna Haraway invites critical antiracist feminists to explore new engagements with the impact of human numbers, despite the fear that doing so opens the door to racist, classist, modernist, and imperialist narratives. She argues that simply acknowledging the urgency of population size and growth offers no specific suggestion for exactly how to address the impacts of human numbers, but she proposes new ways of thinking about reproduction, autonomy, and kinship to do so. Insisting that reproductive freedom "cannot be just a humanist affair, no matter how anti-imperialist, antiracist, anticlassist, and pro-woman,"[2] Haraway envisions new possibilities for thinking about and forming kinship bonds as a way to both celebrate low birth rates and personal decision-making, while flourishing in relationships of caretaking. This would be focused on proliferating "other-than-natal kin,"[3] particularly in wealthy, high consuming countries. In a related vein, indigenous feminist scholar Kim TallBear's work on decolonial love and kin provides some possibilities for rethinking kinship. TallBear argues for reclaiming indigenous ethical relations in ways that de-center the colonizing relationship forms—including monogamy, marriage, private property ownership and nuclear families—that have gone hand in hand with unequal gender relations, toxic relationships with nonhuman animals and nature, and the economic and moral pressure to reproduce the nuclear family unit.

These framings offer important correctives to the limiting narratives that restrict women's roles to reproductive ones, and legitimate families to those defined by blood and lineage. Yet, they are still somewhat vexing. They don't offer a way out of the focus on numbers that grounds understandings of population-environmental problems in biological processes. As this book hopefully demonstrates, direct environmental impacts driven by human numbers are nearly impossible to tease out because they are not, and never have been, simply biological—they are the result of biological, and political, and economic, and technological, and cultural processes and practices. We must contextualize populationism's presents and its futures through these lenses.

I wrote this book because my own experience with the issue of fertility and childbearing in the Global South was so different from what I was hearing from populationists. As an undergraduate participant in a semester abroad program in Kenya, and later as a Peace Corps Volunteer in Madagascar, I spent a significant amount of time in African hospital labor and delivery wards, watching babies being born. Babies, babies everywhere, delivered in all kinds of circumstances: normal, uneventful births, breech births, still births, spontaneous miscarriages, precarious births that happened prematurely. Most of the women who delivered these babies were living in poverty, whether relative poverty or grinding poverty on a scale that most readers of this book can't imagine. During my encounters with women, I often asked about their babies, and what they meant to them. The answers were diverse and multifaceted: they told me that babies were sources of joy, of familial and cultural continuity. For some of the women, babies would solidify their position within new marriages, satisfying husbands' and in-laws' expectations. For many, they would solidify their position within the community as mothers, an important status to hold among other women. And for most, babies were buffers against future economic precarity. Although they began as extra mouths to feed, they would later grow to contribute to family well-being, whether as extra pairs of hands to work the farm, or later, if they were lucky, they would find good office jobs and bring home nice salaries that would help their parents and extended family members. Above all, babies symbolized their hopes for the future.

At the same time, many of the women and adolescents I spoke to preferred to have fewer babies than their parents had. In Kenya, the young women I met were more interested in office jobs, cars, and cell phones than in having large families. The young women I spent my days with in Madagascar rejected a well-known wedding blessing ("May you have seven sons and seven daughters"), dreaming instead of having two or three children and more access to disposable incomes. Adolescent girls in the small, rural Malagasy town where I lived prioritized school and friends, boys, sex, and their own futures. They were interested in becoming mothers, but at a later time, after completing school and establishing households, and also securing incomes. They wanted to leave their small town and see more of the country, if not the world. They wanted bold, bright futures.

I recount these encounters because, when I later participated in population advocacy trainings, I found that the young African women I had spent so much time with were the kinds of women who populationist advocates described through the dualistic lens of victim/agent. The full richness and contexts of their lives were reduced to statistics, quantified abstractions far removed from the complexity and robustness of fertility, reproduction, and childbearing in everyday life. They were described as victims of poverty and limited choices, of regressive cultural traditions and gender norms, and of the effects of soil erosion, storm surges, and heat waves. Yet, they were also described as agents possessing the capacity to complete their educations, earn their own incomes, and consistently use contraceptives. This was development-oriented agency, the kind that reflects a Western model predicated on modernization and neoliberal economic improvement. It was the kind of agency that would position them to limit their reproduction, and to save the world.

This development-led agency is particularly common in books and development programs focused on investing in women and girls. The Girl Effect, an independent NGO initially created by the Nike Foundation, describes the model's ethos on its website:

> When a girl has self-belief and is supported by her family and community; when she's empowered with skills, ideas and knowledge; when she has access to services, role models and other girls; when she is visible and vocal—she can demand to stay in school, to get healthcare, and to get married and have children when she chooses.[4]

The girl described in this model is similar to the sexual steward: a self-regulating, rational and responsible actor who makes decisions that are in line with social and economic improvement, along the lines of development's modernizing rationale. And the tools required to get her there operate through an atomized, individual improvement model aimed at building her self esteem, knowledge, and self-worth through technical interventions. This neoliberal model of individual self-improvement has long circulated through gender and development discourses, grounding both arguments of victimhood and agency, and isolating women's circumstances from political and cultural context.

The "girl" described in the Nike Foundation campaign is also an ideal-type development figure. She first emerged with startling clarity in a 1992 speech by economist Larry Summers as a source of human capital, with endless potential to provide maximum social and economic benefits to her family, community, and society—all facilitated through investments in her formal education. Moreover, she would be able to turn the vicious cycle of endemic poverty into a virtuous circle of education and health for all:

> She has a greater value outside the home and thus has an entirely different set of choices than she would without education. She is married at a later age and is able to better influence family decisions. She has fewer, healthier children and can insist on the development of all of them, ensuring that her daughters are given a fair chance. And the education of her daughters makes it much more likely that the next generation of girls, as well as of boys, will be educated and healthy as well. The vicious cycle is thus transformed into a virtuous circle.[5]

Ironically, "the girl" was also an ideal-type population figure. Summers's speech went on to describe additional gains that could be achieved through investments in girls' education—specifically, reductions in birth rates. He portrayed educating girls as a missing component of economic development for nations in the Global South, in that it could help delay marriage, leading to slower birthrates. Calculating that each year of formal education reduces future fertility rates by 5–10%, Summers concluded that investing in girls through education is less expensive and provides better returns on investment than family planning programs, because of additional benefits such as lower mortality, higher future wages, and higher GDP per capita.

Here, girls replace women as the site for investing in family planning: education is less expensive and provides greater "return on investment" by way of fertility reduction, longevity, and higher wages. Girls are thus positioned as sources of human capital, capable of providing future economic benefits to their families, societies, and nations. As sources of human capital, their rates of return are endlessly beneficial; their bodies become the location for, as Murphy describes it, "an anticipatory, future-oriented calculation of value."[6]

However, as Wilson points out, this "girl" is also a symbol of neoliberal calls for the intensification of female labor. Not only do these kinds of models of so-called empowerment avoid questioning the gendered division of labor, they build on it further by suggesting that investing in girls' potential future labor yields economic and other dividends for family, community, and society at large. Neoliberal economic development is directly implicated in the deep local and regional inequalities that give rise to global poverty, climate change, and population growth.[7] Nevertheless, discourses describing development figures like that of the girl and the sexual steward are continually produced anew. In fact, in 2017, shortly before this writing, a new book was released that described its conclusions as "the most comprehensive plan ever proposed to reverse global warming."[8] Drawing on quantitative data produced by seventy researchers in twenty-two countries, the book listed one hundred solutions to climate change; items #6 and #7 in the list are "family planning" and "educating girls." When researchers combined their effects into a single measure and recalculated, the combined family planning-education solution ranked at #1. Predictably, the book became a bestseller.

It makes sense to search for simple solutions to complex problems. And to be clear, accessing education and contraceptives is not always easy for women and girls around the world. However, the simplicity of instrumental solutions and technical quick fixes prevents researchers and policymakers, as well as activists, from truly understanding the full contexts of the problems they seek to address and finding strategies to address them that will undo unequal social systems and structures, rather than reproduce their effects.

This book has explored the various ways in which population-environment linkages have been articulated by environmental advocates, paying particular attention to developments in scientific knowledge, reproductive politics, and shifts in funding priorities. A central theme operating throughout is that science and political change do not operate in isolation from each other, progressing along separate, parallel tracks. Rather, they develop together, simultaneously, through overlapping entanglements and interactions of actors, funding resources, personal priorities and institutional agendas. In a context of diminishing resources overall, funding for population-environment science has increased, leading to more science-based advocacy.

Despite significant increases in public attention to and writing on population and climate change, population advocates often feel beleaguered by what they view as an unreceptive American public. My research informants often described situations in which they were publicly challenged, often with open hostility, by people who refused to believe that global population should be addressed in any way. As a critical researcher, I experienced just the opposite, particularly during an academic conference when an irate ecology professor in the audience shouted at me, arguing that my critiques of the population-pressure-on-resources model promoted "dangerous" ideas. Regardless of the position, the relationship between population trends and environmental change remains a highly contentious issue. Expressing any opinion on global population issues invites potential backlash from any range of communities, depending on your stance. When I first began researching this topic in graduate school, I discussed it with a fellow academic, who shifted uncomfortably in her seat as soon as I began speaking. "We're not supposed to talk about population . . . are we?" she asked. Apparently not, if we express views that are contrary to the perceived stance of our interlocutor. Unless, that is, we ground our arguments in science.

As I argued in chapters 1–3, scientific knowledge is widely perceived as a means of finding neutral support for ideas about population-environment linkages that are otherwise viewed as political or moral, and thus controversial. With a stronger scientific grounding, population advocates find broader perceptions of legitimacy in their messaging work. In turn, this facilitates the circulation of frameworks linking population reductions and environmental sustainability through the dual lenses of scientific knowledge and women's empowerment. At the same time, the current focus on climate change within the broader environmental and scientific communities, along with the proliferation of scientific studies and models projecting both greenhouse gas emissions and demographic population growth trends, provide advocates with useful tools for framing these linkages as scientifically innovative, urgent, and apolitical.

Increasing scientific knowledge linking population growth with future climate change also facilitates the development of anticipatory politics strategies. Through these strategies, projection models of what could potentially happen forty or ninety years from now are dynamically brought into the present through advocacy messaging, strategically crafted to focus on the dystopian environmental outcomes that would result if foreign policy changes were not made. However, advocacy in this context is highly dynamic and context-driven. Depending on whether the setting is a population-environment advocacy training or campus-based workshop, a research briefing, or a high level funders' network meeting, an advocate can be anyone from a scientist to a donor, a college student, or a long time congressional lobbyist. Population advocates occupy all of these positions and more, as they frequently shift between institutional homes and positions within a broader network. Over the course of this project, the emergent dynamics of shifting network politics were less surprising than the multiplicity of forms advocacy takes, particularly from powerful actors. As chapter 3 demonstrates, the power wielded by private donors advocating for population-environment science and policy interventions from behind the scenes emerged as one of the most important influences on contemporary strategies in this arena. Private donors draw on individual priorities, use personal networks, and bear on their ability to direct large sums of capital to set the agendas of grantees, including scientists and NGO program managers. As a result, the shifting scientific and political discourses in this arena can often be traced to the private agendas of philanthropic organizations whose work is unimpeded by public accountability or democratic participation.

Over the past four decades, environmental NGOs in the U.S. have ridden the waves of fluctuations in funding, corporate opposition to environmental regulation, and disagreements over policy issues within their own membership. Those organizations that have also made forays into population programming, through advocacy, research, or program implementation overseas, have seen a concentration of resistance and waning financial support. They have also seen strong ambivalence among both leadership and their membership bases, based on an aversion to involvement in controversial issues. As my interviews revealed, fear of being associated with abortion proponents, concern over a perceived sense of drift from organizational missions and priorities, and the harsh realities of generally reduced funding for population work have led many environmental organizations to abandon their previously operating population programs. The issue of race and controversy in environmental and population debates has also played an important and contradictory role, both pushing some organizations away from addressing population, as well as inspiring efforts among others to shift the terms of the debate by drawing on social and reproductive justice frameworks.

Some population advocates have identified the social movement frameworks designed to respond to race and gender inequalities—like reproductive justice—as opportune for their own projects. In the contemporary moment, when young activists are concerned with doing socially just work, and donors are interested in supporting reproductive justice and environmental justice projects, it makes sense that populationists would join the bandwagon. But this raises further questions. Social justice takes on a deeper sense of importance in this context, even as its increasing invocation appears to be emptied of transformative critique. When young population advocates hand out flyers and condoms, asking potential supporters to join their advocacy efforts to fight for justice, it is confusing. What social justice is this? Are condoms and birth control pills examples of social justice technologies if nothing shifts in the social worlds of the people using them? What about family planning services? Are they socially just, in and of themselves, if they don't include comprehensive maternal health care, fertility services, STI prevention and treatment, and comprehensive sexual health education (not primarily focused on reproduction)? Where is the *social* in social justice? Who is it for?

In the advocacy arena, messaging is everything. In order to grapple with the historical roots and future legacy of their work, populationists must deal with the racial controversies that have operated throughout much population thinking and population control policies. However, dealing with the racial legacy of populationism is a difficult thing to do. It is easier to adopt the language of a progressive framework and move forward, rather than mine the critical depths of structural injustices and the role organizational interventions play in maintaining those structures. It is harder still to transform the conditions within which their particular corner of development work is located.

While the future of population advocacy is uncertain, transnational youth organizing continues to extend in new directions. In efforts that build and expand on existing organizational frameworks and advocacy approaches, individual activist youth are making their own population-environment advocacy networks, drawing on social media and other digital technology as a basis for transnational organizing. Young people in climate change conferences, drawing on SRHR messages, climate data, and a sense of shared global responsibility in producing a better world, assert their leadership in putting forth the idea that populationism can help support climate change mitigation and adaptation. While their presentations are often fairly simplistic, they demonstrate the transnational reach of women's empowerment and social justice messages as a linchpin of their messaging. These messages are, after all, in line with the girl-centered approach of the future.

Youth advocacy in this arena is still freighted with inequalities in power, privilege, and access. Regardless of whether emergent youth networks are transnational, it is still those youth who can harness the power of development paradigms, networks, and advocacy spaces, that set the terms of the agenda—and as a result, those who could potentially transform this arena of policy advocacy are those who hold the tools of privilege. Sadly, for all of their vibrant enthusiasm, the emergent youth voices I heard at the COY6 meetings in Cancun articulated a narrow population advocacy perspective which ignored global structural inequalities, military- and corporate-led environmental degradation, corporate resource extraction, and the increasing entrenchment of poverty for women and families around the world. Population advocates, whether youth or seasoned activists, in the U.S. or other hemispheres of the world, consistently avoid articulating these inequalities, because doing so does not lend itself to simple, policy-relevant solutions. As a result, their efforts will continue to search in vain for solutions that avoid addressing deep-rooted structural forces, thus keeping us on a business-as-usual development track and, sadly, foreclosing the alternative futures they claim to seek.

ACKNOWLEDGEMENTS

This book has benefitted from the graciousness and generosity of so many people. My first and deepest thanks to those members of the population network who invited me into their offices, boardrooms, and other training and workshop spaces, as well as those who shared their files, reports, and other documents with me. Thank you to my editors at NYU Press, particularly Maryam Arain, for infinite patience and good humor. Thank you to the UC Berkeley faculty who nurtured this project from its infancy, particularly Nancy Peluso, Louise Fortmann, Carolyn Finney, and Lawrence Cohen. Special thanks are also due to Adele Clarke, Michael Watts, Nancy Scheper-Hughes, and Kim TallBear. To my first writing group—Martine Lappe, Rachel Washburn, Katie Hasson, and Theresa MacPhail—who read nearly every early word and helped me make it better, thank you. I am most grateful for the intellectual community and friendship of James Battle, Larisa Kurtovic, Andrew Hao, E. Mara Green, Marcus Moore, Alysoun Quinby, Elizabeth Farfan-Santos, Mez Baker, Megan Ybarra, Jason Morris Jung, Bhavna Shamasunder, and Hodari Toure. To the broader Bay Area community that nurtured me as I nurtured this project, my infinite thanks: Amy Saxton, Michael Dailey, Jill Bates-Moore, Karen Moore, and Jason Woodside.

I am grateful for the financial support of the National Science Foundation and the Ford Foundation, which made this research project possible. I am also deeply grateful to the rich, robust community of scholars, activists, and activist-scholars whose work has inspired and broadened and deepened my thinking over the years. Betsy Hartmann, Banu Subramaniam, Marlene Gerber Fried, Loretta Ross, Rickie Solinger, Michelle Murphy, and Saul Halfon, thank you for your work. Thank you to Anne Hendrixson, Rajani Bhatia, and Ellen Foley for the collaborative thinking and the pushes in the right direction when needed. Much gratitude is owed to my former LMU colleagues, Stella Oh, Traci Voyles, Linh Hua, and Carla Bittel. Many, many thanks to my UCR writing group—

Chikako Takeshita, Dana Simmons, and Juliet McMullin—for your close readings and thoughtful comments. I am grateful to the UCR Science Studies Colloquium for the thoughtful engagement with what would become chapter 4. Thanks also to my department colleagues, Jane Ward, Sherine Hafez, Chikako Takeshita, Margie Waller, Juliann Alison, Katja Guenther, Eric Stanley, Tamara Ho, Amalia Cabezas, Alicia Arrizon, Anthonia Kalu, and Crystal Baik, as well as to Donatella Galella, Ademide Adeluyi-Adelusi, Jody Benjamin, and Derick Fay. I am also immensely grateful to my students, past and present, who have always taught me so much.

Art sustained me, through to the end. The final months of this project were spent on the grounds of one of the most beautiful places I've ever been: the Huntington Library and Gardens. Having a readership there gave me writing space and access to the botanical gardens and art galleries; those months were infinitely beneficial intellectually, emotionally, and aesthetically. Thanks also to the musical artists who provided the soundtrack that accompanied every word, particularly those on the James Brown, Isley Brothers, and Alice Smith stations of my favorite music streaming app.

Infinite thanks to my friends and SGI family, who endured hearing the endless narrative of this book with patience, warmth, and humor. My very deepest appreciation to my Sasser and McCartha families for being the best cheerleaders. Uncle Emile, I wish you were here to see it come to fruition. And finally, to the Sasser four (plus Deanna), this is for you.

NOTES

INTRODUCTION

1 Angus and Butler, *Too Many People?*

2 Sasser, "Population, Climate Change, and the Embodiment of Environmental Crisis."

3 Engelman, *Population, Climate Change, and Women's Lives*, 5.

4 Ibid., 14.

5 Unnamed author, "COP16 Policy Statement," accessed December 1, 2016.

6 Demographic transition theory refers to a four-stage model of transition from high birth and high death rates, to low birth rates and low death rates. It is a central theory within the field of demography, and has been at the heart of international family planning interventions. See chapter 2 for a fuller explanation.

7 Lee and Mason, "What is the Demographic Dividend?," accessed November 24, 2017.

8 Bloom, Canning, and Sevilla, *The Demographic Dividend*, xi.

9 Hendrixson, "Beyond Bonus or Bomb."

10 Dalrymple, "Is the Tea Party a 'Social Justice' Movement?" accessed May 26, 2017, at www.patheos.com; for a discussion of the "social justice for fetuses" argument, see Christopher Kaczor's 2011 book *The Ethics of Abortion: Women's Rights, Human Life, and the Question of Justice*. New York: Routledge.

11 World Bank Open Data, 2015, accessed at www.data.worldbank.org; Union of Concerned Scientists, accessed at ucsusa.org; Global Carbon Atlas, accessed May 26, 2017 at www.globalcarbonatlas.org.

12 United Nations, World Population Prospects, 2015 Revision, accessed May 26, 2017.

13 Unmet need is a concept that arises from demographic surveys and is a central concept in the women's empowerment framework as applied to population policy and reproductive decision-making. It is a measure of the number of women who are sexually active and in a stable union, who state that they are not currently interested in becoming pregnant, but who are not consistently using Western methods of contraception. Halfon (2007) describes unmet need as both a "technical object and a policy pronouncement" (156), closely linked to the notion of "latent demand." In fact, USAID developed demographic surveys partly for this very purpose—to have an instrumental measure that would demonstrate an existing latent demand that could then be translated into an unmet need. Unmet

need is not solely scientific, it is political; policymakers have used it to appeal to a quantified, scientific version of latent demand. Unmet need is central to justifying population programs, and "given the centralized regime of data gathering, unmet need discourse helps to maintain a few key institutions as the locus of program expertise and development" (157).

14 United Nations, World Population Prospects, 2015 Revision, accessed May 26, 2017; Tabutin and Schoumaker, "The Demography of Sub- Saharan Africa from the 1950s to the 2000s: A Survey of Changes and a Statistical Assessment"; United Nations, 2003, World Population Prospects, The 2002 Revision (New York: United Nations).
15 Ferguson, *Global Shadows*, 2.
16 Adichie, Chimamanda Ngozi, 2009, TEDGlobal, www.ted.com/talks.
17 Dogra, "The Mixed Metaphor of 'Third World Woman.'"
18 Win, "Not Very Poor, Powerless, or Pregnant," 79.
19 Gilens, "How the Poor Became Black."
20 Dogra, "The Mixed Metaphor of 'Third World Woman.'"
21 Mohanty, "Under Western Eyes," 56.
22 Robbins, *Political Ecology.*
23 Blaikie and Brookfield, Harold, *Land Degradation and Society.*
24 Harvey, "Population, Resources, and the Ideology of Science."
25 Watts, *Silent Violence*; Davis, *Late Victorian Holocausts.*
26 Davis, *Late Victorian Holocausts.*
27 Fairhead and Leach, *Misreading the African Landscape.*
28 Jarosz, "Defining Tropical Deforestation."
29 Hartmann, *Reproductive Rights and Wrongs*; Mamdani, *The Myth of Population Control.*
30 Hajer, "Discourse Coalitions," 45.
31 Ibid.
32 Keck and Sikkink, "Transnational Advocacy Networks."
33 Agarwal, *Gender and Green Governance*; Rocheleau, *Feminist Political Ecology.*
34 MacGregor, "'Gender and Climate Change.'"
35 Leach, "Earth Mother Myths and Other Ecofeminist Fables"; Resurreccion, "Persistent Women and Environment Linkages."
36 Ibid.
37 Braidotti, Charkiewicz, Hausler, and Wieringa, *Women, the Environment and Sustainable Development*, 78.
38 Ibid.
39 Rathgeber, "WID, WAD, GAD."
40 Buvinic, "Projects for Women in The Third World."
41 Moser, *Gender Planning and Development*, 61.
42 Kabeer, *Reversed Realities.*
43 Ibid., 224–25.
44 Sen and Batliwala, "Empowering Women for Reproductive Rights," 17.

45 Sen and Grown, *Development, Crises, and Alternative Visions*, 19.
46 Ibid.
47 Halfon, *The Cairo Consensus*.
48 Ibid., 96.
49 Kabeer, *Reversed Realities*, 224.
50 Foucault, *The History of Sexuality*.
51 Adams and Pigg, *Sex in Development*.
52 Jasanoff, *States of Knowledge*, 17.
53 Ibid., 20.
54 Reardon, *Race to the Finish*, 8.
55 Kaiser Family Foundation, 2017, "The U.S. Government and International Family Planning & Reproductive Health Efforts," www.kff.org.

CHAPTER 1. THE POPULATION "CRISIS" RETURNS

1 The imagery clearly references the opening scene of Paul Ehrlich's book *The Population Bomb*, by far the most well-known population control advocacy book in the U.S. It opens with a famous scene of Ehrlich and his family in a taxi cab in India, nervously navigating the throng of brown bodies that surround them on all sides. For Ehrlich, this is the moment when population moves from being an intellectual issue to a visceral one, based on his family's frightened reaction to the crowd.
2 Whitty, "The Last Taboo," 26.
3 Verilli and Piscitelli, "Research Findings Report."
4 Engelman, "Population & Sustainability."
5 See "U.N. Forecasts 10.1 Billion People by Century's End" and associated blog posts: www.nytimes.com.
6 There are many articles on this theme, which tend to proliferate in November and December, around the time of the IPCC climate change conference. Examples include: Emmott, "Though Climate Change is a Crisis," accessed November 1, 2016 ; *Los Angeles Times* Editorial Board, "Why We Need to Address Population Growth's Effects on Global Warming," accessed November 1, 2016 ; Plautz, "The Climate-Change Solution No One Will Talk About," accessed November 1, 2016 ; Zelman, "World Population Day," accessed November 1, 2016.
7 *Los Angeles Times* Editorial Board. "Why We Need to Address Population Growth's Effects on Global Warming."
8 Prois, "Voluntary Birth Control."
9 IPCC, *Climate Change 2014 Synthesis Report Summary for Policymakers*, 5.
10 Plautz, "The Climate-Change Solution No One Will Talk About," accessed April 1, 2015; Goering, "Family Planning 'Effective' but Unpopular Climate Change Solution," accessed May 20, 2015.
11 Merchant, *Prediction and Control*.
12 Connelly, *Fatal Misconception*.
13 Ibid.

14 Ibid.; Hartmann, *Reproductive Rights & Wrongs*; Sullivan, *Leveraging the Global Agenda for Progress*.
15 Kabeer, *Reversed Realities*, 189.
16 Connelly, *Fatal Misconception*, 355.
17 Cock, "The World Women's Congress."
18 Various authors, *Treaty on Population, Environment and Development*, paragraph 1, accessed June 2017.
19 Ibid., paragraph 3.
20 Ibid., 7.
21 Petchesky, *Global Prescriptions*.
22 United Nations, Programme of Action.
23 Ibid., 64.
24 Hodgson and Cotts Watkins, "Feminists & Neo-Malthusians."
25 Campbell, "Schools of Thought."
26 Sasser, "Environmental Organizations and Population Programs."
27 McCoy, Chand, and Sridhar, "Global Health Funding."
28 Foley and Hendrixson, "From Population Control to AIDS."
29 Guttmacher Institute, *Just the Numbers*, accessed June 2017 at www.guttmacher.org.
30 Anonymous interview. Former staff member at USAID Office of Population and Reproductive Health. In-person interview, San Francisco, CA. 2010.
31 Anonymous interview. Former staff member of the Compton Foundation. In-person interview, Menlo Park, CA. 2010.
32 Costello, et al., "Managing the Health Effects of Climate Change."
33 McKibben, "Global Warming's Terrifying New Math," accessed June 10, 2016.
34 McKibben, "A World at War," accessed October 2, 2016.
35 Peet, Robbins, and Watts, *Global Political Ecology*.
36 Hickey, Rieder, and Earl, "Population Engineering," 845.
37 Ibid., 855.
38 Ibid., 866.
39 Hartmann, *The America Syndrome*.
40 Fuller, "Demographics = Mideast Turmoil."
41 Quoted in Hartmann and Hendrixson, "Pernicious Peasants and Angry Young Men."
42 Leahy, Engelman, Vogel, Haddock, and Preston, *The Shape of Things to Come*, 8.
43 Ibid.
44 Hartmann and Hendrixson, "Pernicious Peasants and Angry Young Men."
45 Potts, "Crisis in the Sahel," 3.
46 Ibid., 4.

CHAPTER 2. HOW POPULATION BECAME AN ENVIRONMENTAL PROBLEM

1 Jowett, "Plato on Population and the State."
2 Federici, *Caliban and the Witch*.

3 Hutchinson, *The Population Debate.*
4 Malthus, *An Essay on the Principle of Population*, 12.
5 Darwin, "Darwin to Wallace, 1858," 465.
6 Galton, *Inquiries into Human Faculty and its Development.*
7 Ibid., 17.
8 Galton, "The Possible Improvement of the Human Breed."
9 Stern, *Eugenic Nation*, 135.
10 Ibid.; Wohlforth, "Conservation and Eugenics."
11 Gottlieb, *Forcing the Spring.*
12 Warren, *Aldo Leopold's Legacy*, 33.
13 Roosevelt, "Opening Address by the President," 3.
14 As quoted in a book of essays written by two sisters for *Everybody's Magazine*, chronicling their experiences traveling the country and performing "women's work." Roosevelt wrote the letter to the sisters. Van Vorst and Van Vorst, *The Woman Who Toils.*
15 Roosevelt, "On American Motherhood."
16 Merchant, "Shades of Darkness," 382–3.
17 Fisher, *Report on National Vitality, Its Wastes and Conservation*, 55.
18 Ibid., 129.
19 Stern, *Eugenic Nation.*
20 Greene, *Malthusian Worlds*, 3.
21 Ibid., 30.
22 Hodgson, "The Ideological Origins."
23 Ibid., 5.
24 Ibid., 18.
25 Ibid., 21.
26 Notestein, "Population-The Long View."
27 Szreter, "The Idea of Demographic Transition," 660.
28 Ibid., 662.
29 Demeny, "Social Science and Population Policy," 455.
30 Hodgson, "Demography as Social Science and Policy Science."
31 Sharpless, "Population Science, Private Foundations, and Development Aid," 176.
32 Hodgson, "Demography as Social Science and Policy Science."
33 Briggs, *Reproducing Empire*, 8.
34 McCann, *Figuring the Population Bomb.*
35 Ibid.
36 Porter, *Trust in Numbers*, ix.
37 Ibid.
38 Briggs, *Reproducing Empire*, 82.
39 Pearl, *The Biology of Population Growth*; Pearl, "The Growth of Populations."
40 Pearl, *The Biology of Population Growth*, 18.
41 Pearl, "The Population Problem," 640.
42 Höhler, "A 'Law of Growth.'"

43 Pearl, *The Natural History of Population*, 2.
44 Kingsland, "The Refractory Model"; Sayre, "The Genesis, History, and Limits of Carrying Capacity."
45 Bashford, *Global Population*.
46 Ibid., 6–7.
47 Höhler, *Spaceship Earth in the Environmental Age*.
48 Leopold, "Ecology and Politics," 282.
49 Ibid., 284.
50 Sayre, "The Genesis, History, and Limits of Carrying Capacity," 132.
51 Vogt, *Road to Survival*, 14.
52 Ibid., 16.
53 Ibid., 17.
54 Ibid., 282.
55 Connelly, *Fatal Misconception*.
56 Osborn, *The Limits of the Earth*.
57 Robertson, *The Malthusian Moment*.
58 Ibid., 8.
59 Hardin, "The Tragedy of the Commons," 1244.
60 Ibid.
61 Robertson, The Malthusian Moment, 153.
62 Hardin, "The Tragedy of the Commons," 1248.
63 Ibid., 1247.
64 Hardin, "Commentary," 562.
65 Hardin, "Parenthood: Right or Privilege?," 427.
66 Robertson, *The Malthusian Moment*, 191.
67 Ehrlich, *The Population Bomb*, 1.
68 Ibid., xii.
69 Ibid., 1.
70 Ehrlich and Harriman, *How to be a Survivor*, 23.
71 Ibid., 23.
72 Ibid., 26.
73 Gottlieb, *Forcing the Spring*.
74 Ehrlich and Holdren, "Impact of Population Growth," 1212.
75 Meadows, Randers, and Meadows, *Limits to Growth*, xxi.

CHAPTER 3. SCIENTISTS, DONORS, AND THE POLITICS OF ANTICIPATING THE FUTURE

1 Adams, Murphy, and Clarke, "Anticipation," 249.
2 Dowie, *American Foundations*; Hartmann, "Strategic Scarcity."
3 O'Neill, MacKellar, and Lutz, *Population and Climate Change*, xiii.
4 O'Neill, Dalton, Fuchs, Jiang, Pachauri, and Zigova, "Global Demographic Trends," 5.
5 Murtaugh and Schlax, "Reproduction and the Carbon Legacies of Individuals."

6 Satterthwaite, “The Implications of Population Growth.”
7 Tapia Granados, Ionides, and Carpintero, “Climate Change and the World Economy.”
8 Martine, “Population Dynamics and Policies.”
9 Population Action International originated in the same year as Hugh Moore and William Draper’s Population Crisis Committee (see chapter 1). Today it describes itself as a “global organization advancing the right to affordable, quality contraception and reproductive health care for every woman, everywhere.” Population Action International website, “Who We Are,” www.pai.org.
10 Mutunga and Hardee, “Population and Reproductive Health,” 189.
11 The project, and PAI’s climate-related portfolio, has since been discontinued.
12 Mutunga, Zulu, and DeSouza, “Population Dynamics, Climate Change, and Sustainable Development in Africa,” accessed June 2017.
13 Ibid., 17.
14 Hartmann, “Converging on Disaster.”
15 Hartmann, “Liberal Ends, Illiberal Means.”
16 IPCC Data Distribution Center, “SRES Emissions Scenarios,” accessed April 29, 2017.
17 Nakicenovic et al., *Special Report on Emissions Scenarios*, 599.
18 McCracken, “Prediction Versus Projection-Forecast Versus Possibility,” accessed March 15, 17.
19 Website previously located at www.popoffsets.org. Accessed March 2016.
20 Bumpus and Liverman, “Carbon Colonialism?,” 204–5.
21 Wire, “Fewer Emitters, Lower Emissions, Less Cost.”
22 Murphy, *The Economization of Life*, 6.
23 Hull, “China Right to Link Population to Climate.”
24 The one-child policy, which formed the national population planning policy in China, was formally instituted in 1979, and began to be phased out in 2015. The policy stipulated that urban couples in densely populated areas have no more than one child, although exceptions were provided for ethnic minorities and parents whose first child had physical or cognitive disabilities, or if the first child was a girl.
25 Murphy, *The Economization of Life*.
26 All names and initials in this book, other than the author’s, are pseudonymous.
27 Kincaid, Dupré, and Wylie, *Value-Free Science?*
28 Wilson and Kehoe, “Environmental Organizations”; Pielemeier, “Review of Population-Health-Environment Programs”; Kleinau, Randriamananjara, and Rosensweig, “Healthy People in a Healthy Environment.”
29 Greenhalgh, *Just One Child*, 8.
30 Ibid.
31 Dowie, *American Foundations*.
32 Sasser, “Environmental Organizations and Population Programs.”
33 Pielemeier, “Review of Population-Health-Environment Programs.”

34 Speidel et al., "Making the Case."

CHAPTER 4. THE ROLE OF YOUTH IN POPULATION-ENVIRONMENT ADVOCACY

1 GPEP has since been phased out; the Sierra Club has subsumed its reproductive health work under a new program called the Gender, Equity, and Environment program, which focuses on climate policy, health, and energy access that prioritize women, transgender, and gender nonconforming people. The new program has a much smaller focus on SRHR and family planning, but still maintains a webpage on "Contraception and Climate Change," located at sierraclub.org.

2 Malkki, *The Need to Help*, 4.

3 Ibid., 8.

4 Greenbaum, "'Sex and Sustainability' Event."

5 Resurreccion, "Persistent Women and Environment Linkages."

6 This was before the Black Lives Matter movement exploded onto the American social movement scene in 2013.

7 Population Connection, "30 Years of ZPG: Our History as Written in 1998," accessed June 2017 at www.populationconnection.org.

8 *LIFE* Magazine, "ZPG."

9 ZPG Reporter, "ZPG Goals," 2.

10 As quoted in Robertson, *The Malthusian Moment*, 160.

11 Ibid., 160.

12 Ibid.

13 ZPG Reporter, "ZPG Goals."

14 Examples include: "When you're feeling tender, think about the Hellbender," "Wrap with Care . . . Save the Polar Bear," and "Fumbling in the Dark? Think of the Monarch." See www.endangeredspeciescondoms.com for more.

15 Sasser, "From Darkness Into Light."

16 Mazur, *A Pivotal Moment*.

17 Ibid., 1.

18 Ibid., 17. Also see chapter 5 for a critical discussion of how populationists draw on the reproductive justice framework.

19 Ibid.

20 Ibid., 18.

21 Reisch, "Defining Social Justice," 343.

22 In the 1990s, Sierra Club members attempted to take over club leadership and institute neo-Malthusian anti-immigration policy advocacy within the organization. This led to bitter disputes between the members, and raised significant questions about the role of racism in majority-white environmental organizations, particularly those that do population work. See Sasser, "From Darkness Into Light."

23 Unnamed author, "COP16 Policy Statement," accessed January 1, 2017.

CHAPTER 5. CO-OPTING REPRODUCTIVE JUSTICE

1 Luna, "From Rights to Justice."
2 Silliman, Fried, Ross, and Gutierrez, *Undivided Rights*; Price, "What is Reproductive Justice?"
3 Silliman and King, *Dangerous Intersections.*
4 Cornwall and Brock, "What do Buzzwords do for Development Policy?," 1.
5 Ibid., 4.
6 Intersectionality is a term coined by legal scholar Kimberlé Crenshaw to describe the ways social identities intersect and overlap in systems of domination or oppression. It is an explanatory framework for understanding social injustice based on multiple aspects of identity and categories of social difference, including race, gender, class, and sexual orientation, among others. See Crenshaw, "Mapping the Margins."
7 Cornwall and Brock, "What do Buzzwords do for Development Policy?," 1.
8 Sasser, "From Darkness Into Light."
9 Davis, *Women, Race, & Class*; Roberts, *Killing the Black Body.*
10 Davis, *Women, Race, & Class.*
11 Solinger, *Pregnancy and Power.*
12 Roberts, *Killing the Black Body*, 4.
13 Ibid., 3.
14 Davis, *Women, Race, & Class.*
15 Silliman, Fried, Ross, and Gutierrez, *Undivided Rights*; Nelson, *Women of Color.*
16 Slater, "Sterilization," 151.
17 Ibid.
18 Stern, *Eugenic Nation.*
19 Dreifus, "Sterilizing the Poor."
20 Ibid., 16.
21 Ibid.
22 Ibid., 8.
23 Lawrence, "The Indian Health Service."
24 Smith, *Conquest.*
25 Ibid., 36.
26 Kluchin, *Fit to Be Tied.*
27 Ibid., 22.
28 Silliman, Fried, Ross, and Gutierrez, *Undivided Rights*, 42.
29 Turshen, *Women's Health Movements.*
30 Luna, "From Rights to Justice."
31 National Organization for Women, "March News-131 Days Until March for Women's Lives!," December 16, 2003.
32 Luna, "From Rights to Justice."
33 Johnson, "Female Inmates Sterilized," accessed May 12, 2017.
34 Ibid.
35 Ibid.

CONCLUSION

1 Hymas, "I Am the Population Problem," accessed June 2017.
2 Haraway, *Staying With the Trouble.*
3 Ibid., 209.
4 The Girl Effect, "Our Purpose," accessed June 2017 at www.girleffect.org.
5 Summers, "Investing in *All* the People."
6 Murphy, *The Economization of Life.*
7 Silliman and King, *Dangerous Intersections*; see also Peet, Robbins, and Watts, *Global Political Ecology.*
8 Hawken, *Drawdown.*

BIBLIOGRAPHY

Adams, Vincanne, and Stacey Leigh Pigg, eds. 2005. *Sex in Development: Science, Sexuality, and Morality in Global Perspective*. Durham, NC: Duke University Press.

Adams, Vincanne, Michelle Murphy, and Adele Clarke. 2009. "Anticipation: Technoscience, Life, Affect, Temporality." *Subjectivity* 28: 246–65.

Agarwal, Bina. 2010. *Gender and Green Governance: The Political Economy of Women's Presence Within and Beyond Community Forestry*. New York: Oxford University Press.

Amalric, Franck. 1994. "Finiteness, Infinity, and Responsibility: The Population-Environment Debate." In Harcourt, Wendy, ed. 1994. *Feminist Perspectives on Sustainable Development*. London: Zed Books.

Angus, Ian, and Simon Butler. 2011. *Too Many People? Population, Immigration, and the Environmental Crisis*. Chicago: Haymarket Books.

Arora-Jonsson, Seema. 2010. "Virtue and Vulnerability: Discourses on Women, Gender and Climate Change." *Global Environmental Change* 21: 744–51.

Bashford, Alison. 2016. *Global Population: History, Geopolitics, and Life on Earth*. New York: Columbia University Press.

Batliwala, Srilatha. 1994. "The Meaning of Women's Empowerment: New Concepts from Action." In Sen, Gita, Adrienne Germain, and Lincoln C. Chen, eds. 1994. *Population Policies Considered*. Cambridge, MA: Harvard University Press.

Bhatia, Rajani, Jade Sasser, Diana Ojeda, Anne Hendrixson, Sarojini Nadimpally, and Ellen Foley. Forthcoming. "Demo-, Geo-, and Bio-Populationism: A Feminist Conceptual Framework on Population in an Era of Climate Change and Unsustainable Development." *Gender, Place, and Culture*.

Bill and Melinda Gates Foundation. 2012. "Innovative Partnership to Deliver Convenient Contraceptives to up to Three Million Women." Accessed June 2017. www.gatesfoundation.org.

Blaikie, Piers, and Harold Brookfield. 1987. *Land Degradation and Society*. London: Methuen.

Bloom, David, David Canning, and Jaypee Sevilla. 2003. *The Demographic Dividend: A New Perspective on the Economic Consequences of Population Change*. Santa Monica, CA: RAND.

Braidotti, Rosi, Ewa Charkiewicz, Sabine Hausler, and Saskia Wieringa. 1994. *Women, the Environment and Sustainable Development: Towards a Theoretical Synthesis*. London: Zed Books.

Briggs, Laura. 2002. *Reproducing Empire: Race, Sex, Science, and U.S. Imperialism in Puerto Rico*. Berkeley: University of California Press.

Bumpus, Adam, and Diana Liverman. 2011. "Carbon Colonialism? Offsets, Greenhouse Gas Reductions, and Sustainable Development." In Peet, Richard, Paul Robbins, and Michael Watts. *Global Political Ecology*. London: Routledge.

Buvinic, Mayra. 1986. "Projects for Women in The Third World: Explaining Their Misbehavior." *World Development* 14(5): 653–64.

Campbell, Martha. 1998. "Schools of Thought: An Analysis of Interest Groups Influential in International Population Policy." *Population and Environment* 19(6): 487–512.

Cock, Jacklyn. 1992. "The World Women's Congress for a Healthy Planet." *Agenda: Empowering Women for Gender Equity* 12: 63–66.

Connelly, Matthew. 2008. *Fatal Misconception: The Struggle to Control World Population*. Cambridge, MA: Harvard University Press.

Cornwall, Andrea, and Karen Brock. 2004. "What do Buzzwords do for Development Policy?: A Critical Look at 'Poverty Reduction,' 'Participation,' and 'Empowerment.'" Prepared for the UNRISD conference on Social Knowledge and International Policy Making: Exploring the Linkages, 20–21 April 2004. Geneva, Switzerland.

Correa, Sonia, and Rosalind Petchesky. 1994. "Reproductive and Sexual Rights: A Feminist Perspective." In Sen, Gita, Adrienne Germain, and Lincoln C. Chen, eds. 1994. *Population Policies Considered*. Cambridge, MA: Harvard University Press.

Correa, Sonia, and Rebecca Reichmann, with DAWN. 1994. *Population and Reproductive Rights: Feminist Perspectives from the South*. London: Zed Books.

Costello, Anthony, et al. 2009. "Managing the Health Effects of Climate Change." *The Lancet* 373: 1693–733.

Crenshaw, Kimberlé W. 1991. "Mapping the Margins: Intersectionality, Identity Politics, and Violence Against Women of Color." *Stanford Law Review* 43(6): 1241–299.

Dabelko, Geoffrey. 2011. "Population and Environment Connections: The Role of U.S. Family Planning Assistance in U.S. Foreign Policy." Working Paper. New York: Council on Foreign Relations.

Dalrymple, Timothy. 2010. "Is the Tea Party a 'Social Justice' Movement?" *Patheos*. www.patheos.com.

Darwin, Charles. 1960 [1858]. "Darwin to Wallace, 1858." In "Darwin's Notebooks on Transmutation of Species." Part II, *Bulletin of the British Museum (Natural History) Historical Series*.

Davis, Angela. 1983. *Women, Race, & Class*. New York: Vintage Books.

Davis, Mike. 2000. *Late Victorian Holocausts: El Nino Famines and the Making of the Third World*. London: Verso Books.

Demeny, Paul. 1988. "Social Science and Population Policy." *Population and Development Review* 14(3): 451–79.

Dogra, Nandita. 2011. "The Mixed Metaphor of 'Third World Woman': Gendered Representations by International Development NGOs." *Third World Quarterly* 32(2): 333–48.

Dowie, Mark. 2001. *American Foundations: An Investigative History.* Cambridge, MA: MIT Press.
Dreifus, Claudia. 1975. "Sterilizing the Poor." *The Progressive* 39: 13–19.
Dyson, Tim. 2005. "On Development, Demography and Climate Change: The End of the World as We Know It?" *Population and Environment* 27(2): 117–49.
Ehrlich, Paul. 1971 [1968]. *The Population Bomb.* Cutchogue, NY: Buccaneer Books.
Ehrlich, Paul, and John Holdren. 1971. "Impact of Population Growth." *Science* 171(3977): 1212–217.
Ehrlich, Paul, and Richard Harriman. 1971. *How to be a Survivor: A Plan to Save Spaceship Earth.* London: Ballantine Books.
Emmott, Stephen. "Though Climate Change is a Crisis, The Population Threat is Even Worse." *The Guardian*, December 4, 2015. www.theguardian.com.
Engelman, Robert. 2009a. Worldwatch Report 183. *Population, Climate Change, and Women's Lives.* Washington, DC: Worldwatch Institute.
———. 2009b. "Population & Sustainability." *Scientific American: Earth 3.0* 19(2): 22–29.
Fairhead, James, and Melissa Leach. 1996. *Misreading the African Landscape: Society and Ecology in a Forest-Savanna Mosaic.* Cambridge: Cambridge University Press.
Federici, Silvia. 2004. *Caliban and the Witch.* New York: Autonomedia.
Ferguson, James. 2006. *Global Shadows: Africa in the Neoliberal World Order.* Durham, NC: Duke University Press.
Fisher, Irving. *Report on National Vitality, Its Wastes and Conservation.* Committee of One Hundred on National Health. Washington, DC: National Conservation Commission. www.us.archive.org.
Foley, Ellen E., and Anne Hendrixson. 2011. "From Population Control to AIDS: Conceptualising and Critiquing the Global Crisis Model." *Global Public Health* 6(sup3): S310–S322.
Foucault, Michel. 1978. *The History of Sexuality: Volume I: An Introduction.* New York: Random House.
Fuller, Graham. 2003. "Demographics = Mideast Turmoil." *Newsday*, September 29, 2003: 21A.
Galton, Francis. 2004 [1883]. *Inquiries into Human Faculty and its Development.* Galton online archives collection. www.galton.org.
———. 1901. "The Possible Improvement of the Human Breed Under the Existing Conditions of Law and Sentiment." *Nature* 64(1670): 659–65.
Gilens, Martin. 2003. "How the Poor Became Black: The Racialization of American Poverty in the Mass Media." In Schram, Sanford F., Joe Soss, and Richard Fording, eds. 2003. *Race and the Politics of Welfare Reform.* Ann Arbor: University of Michigan Press.
Goering, Laurie. 2011. "Family Planning 'Effective' but Unpopular Climate Change Solution." Thomson Reuters Foundation. Accessed May 2015. www.trust.org.
Gottlieb, Robert. 2005. *Forcing the Spring: The Transformation of the American Environmental Movement.* Washington, DC: Island Press.

Greenbaum, Leah. 2009. "'Sex and Sustainability' Event Mixes Art and Development." *The Daily Californian*. Accessed April 29, 2009. http://archive.dailycal.org.

Greene, Ronald Walter. 1999. *Malthusian Worlds: U.S. Leadership and the Governing of the Population Crisis*. Boulder, CO: Westview Press.

Greenhalgh, Susan. 2008. *Just One Child: Science and Policy in Deng's China*. Berkeley, CA: University of California Press. 1996.

———. "The Social Construction of Population Science: An Intellectual, Institutional, and Political History of Twentieth Century Demography." *Comparative Studies in Society and History* 38(1): 26–66.

Gusterson, Hugh. 2008. "Nuclear Futures: Anticipatory Knowledge, Expert Judgment, and the Lack that Cannot be Filled." *Science and Public Policy* 35(8): 551–60.

Guttmacher Institute. 2017. *Just the Numbers: The Impact of U.S. International Family Planning Assistance*. www.guttmacher.org.

Hajer, Maarten A. 1993. "Discourse Coalitions and the Institutionalization of Practice: The Case of Acid Rain in Britain." In Fischer, Frank, and John Forester, eds. 1993. *The Argumentative Turn in Policy Analysis and Planning*. Durham, NC: Duke University Press.

Halfon, Saul. 2007. *The Cairo Consensus: Demographic Surveys, Women's Empowerment, and Regime Change in Population Policy*. Lanham, MD: Lexington Books.

Haraway, Donna. 2016. *Staying with the Trouble: Making Kin in the Chthulucene*. Durham, NC: Duke University Press.

Hardin, Garrett. 1974. "Commentary: Living on a Lifeboat." *BioScience* 24(10): 561–68.

———. 1970. "Parenthood: Right or Privilege?" *Science* 169: 427.

———. 1968. "The Tragedy of the Commons." Science 162(3859): 1243–48.

Hartmann, Betsy. 2017. *The America Syndrome: Apocalypse, War, and Our Call to Greatness*. New York: Seven Stories Press.

———. 2014. "Converging on Disaster: Climate Security and the Malthusian Anticipatory Regime for Africa." *Geopolitics* 19(4): 1–27.

———. 2006. "Liberal Ends, Illiberal Means: National Security, 'Environmental Conflict' and the Making of the Cairo Consensus." *Indian Journal of Gender Studies* 13(2): 195–226.

———. 2002. "Strategic Scarcity: The Origins and Impact of Environmental Conflict Ideas." Unpublished dissertation filed at London School of Economics.

———. 1995 [1987]. *Reproductive Rights and Wrongs: The Global Politics of Population Control*. Cambridge, MA: South End Press.

Hartmann, Betsy, and Anne Hendrixson. 2005. "Pernicious Peasants and Angry Young Men: The Strategic Demography of Threats." In Hartmann, Betsy, Banu Subramaniam, and Charles Zerner. 2005. *Making Threats: Biofears and Environmental Anxieties*. New York: Rowman & Littlefield Publishers.

Harvey, David. 1974. "Population, Resources, and the Ideology of Science." *Economic Geography* 50(3): 256–77.

Hawken, Paul. 2017. *Drawdown: The Most Comprehensive Plan Ever Proposed to Reverse Global Warming*. New York: Penguin Books.

Hendrixson, Anne. 2014. "Beyond Bonus or Bomb: Upholding the Sexual and Reproductive Health of Young People." *Reproductive Health Matters* 22(43): 125–34.

———. 2004. "Angry Young Men, Veiled Young Women: Constructing a New Population Threat." Briefing 34. London: The Corner House.

Hickey, Colin, Travis N. Rieder, and Jake Earl. 2016. "Population Engineering and the Fight Against Climate Change." *Social Theory and Practice* 42(4): 845–70.

Hodgson, Dennis. 1991. "The Ideological Origins of the Population Association of America." *Population and Development Review* 17(1): 1–34.

———. 1983. "Demography as Social Science and Policy Science." *Population and Development Review* 9(1): 1–34.

Hodgson, Dennis, and Susan Cotts Watkins. 1997. "Feminists & Neo-Malthusians: Past and Present Alliances." *Population and Development Review* 23(3): 469–523.

Höhler, Sabine. 2015. *Spaceship Earth in the Environmental Age, 1960–1990*. London: Pickering & Chatto.

———. 2005. "A 'Law of Growth': The Logistic Curve and Population Control Since World War II." Unpublished paper presented at the Technological and Aesthetic (Trans)Formations of Society conference, October 12–14, 2005. Darmstadt Technical University.

Hull, Crispin. 2009. "China Right to Link Population to Climate." www.crispinhull.com.

Human Rights Watch. 2012. "India: Target-Driven Sterilization Harming Women." www.hrw.org.

Hutchinson, Edward Prince. 1967. *The Population Debate: The Development of Conflicting Theories Up to 1900*. Boston: Houghton Mifflin.

Hymas, Lisa. 2011. "I Am the Population Problem." *Rewire*. www.rewire.news.

IPCC: Intergovernmental Panel on Climate Change. 2014. *Climate Change 2014 Synthesis Report Summary for Policymakers*. www.ipcc.ch.

———. 2007. *Climate Change 2007: The Fourth Assessment Report*. www.ipcc.ch.

IPCC Data Distribution Center. n.d. *SRES Emissions Scenarios*. www.sedac.ipcc-data.org.

Jarosz, Lucy. 1993. "Defining Tropical Deforestation: Shifting Cultivation and Population Growth in Colonial Madagascar." *Economic Geography* 69(4): 366–79.

Jasanoff, Sheila. 2005. "In the Democracies of DNA: Ontological Uncertainty and Political Order in Three States." *New Genetics and Society* 24(2): 139–56.

———. 2004. *States of Knowledge: The Co-Production of Science and Social Order*. London: Routledge.

Johnson, Corey G. 2013. "Female Inmates Sterilized in California Prisons Without Approval." Center for Investigative Reporting. www.revealnews.org.

Jowett, Benjamin. 1986. "Plato on Population and the State." *Population and Development Review* 12(4): 781–98.

Kabeer, Naila. 1994. *Reversed Realities: Gender Hierarchies in Development Thought*. London: Verso.

Keck, Margaret, and Kathryn Sikkink. 1999. "Transnational Advocacy Networks in International and Regional Politics." *International Social Science Journal* 51(159): 89–101.

Kevles, Daniel J. 1985. *In the Name of Eugenics: Genetics and the Uses of Human Heredity.* New York: Knopf.

Kincaid, Harold, John Dupré, and Alison Wylie, eds. 2007. *Value-Free Science? Ideals and Illusions.* New York: Oxford University Press.

Kingsland, Sharon. 1982. "The Refractory Model: The Logistic Curve and the History of Population Ecology." *Quarterly Review of Biology* 57(1): 29–52.

Kleinau, Eckhard, Odile Randriamananjara, and Fred Rosensweig. 2005. "Healthy People in a Healthy Environment: Impact of an Integrated Population, Health, and Environment Program in Madagascar." Washington, DC: USAID/Environmental Health Project.

Kluchin, Rebecca M. 2011. *Fit to Be Tied: Sterilization and Reproductive Rights in America, 1950–1980.* New Brunswick, NJ: Rutgers University Press.

Lawrence, Jane. 2000. "The Indian Health Service and the Sterilization of Native American Women." *American Indian Quarterly* 24(3): 400–419.

Leach, Melissa. 2007. "Earth Mother Myths and Other Ecofeminist Fables: How a Strategic Notion Rose and Fell." *Development and Change* 38(1): 67–85.

Leahy, Elizabeth, Robert Engelman, Carolyn Gibb Vogel, Sarah Haddock, and Tod Preston. 2010. *The Shape of Things to Come: Why Age Structure Matters to a Safer, More Equitable World.* Washington, DC: Population Action International.

Lee, Ronald, and Andrew Mason. 2006. "What is the Demographic Dividend?" *Finance and Development: A Quarterly Magazine of the IMF* 43(3). www.imf.org.

Leopold, Aldo. 1941. "Ecology and Politics." In Flader, Susan L. and J. Baird Callicott, eds. 1991. *The River of the Mother of God and Other Essays by Aldo Leopold.* Madison: University of Wisconsin Press.

LIFE Magazine. 1970. "ZPG: A New Movement Challenges the U.S. to Stop Growing." *LIFE* 88(14): 32–37.

Lizarzaburu, Javier. 2015. "Forced Sterilisation Haunts Peruvian Women Decades On." *BBC News*, December 2, 2015. www.bbc.com.

Los Angeles Times Editorial Board. 2015. "Why We Need to Address Population Growth's Effects on Global Warming." January 25, 2015. www.latimes.com.

Luna, Zakiya. 2009. "From Rights to Justice: Women of Color Changing the Face of US Reproductive Rights Organizing." *Societies Without Borders* 4(3): 343–65.

MacGregor, Sherilyn. 2010. "'Gender and Climate Change': From Impacts to Discourses." *Journal of the Indian Ocean Region* 6(2): 223–38.

Malkki, Liisa. 2015. *The Need to Help: The Domestic Arts of International Humanitarianism.* Durham, NC: Duke University Press.

Malthus, Thomas Robert. 2004 [1798]. *An Essay on the Principle of Population.* Oxford: Oxford University Press.

Mamdani, Mahmood. 1972. *The Myth of Population Control: Family, Caste and Class in an Indian Village.* New York: Monthly Review Press.

Martine, George. 2009. "Population Dynamics and Policies in the Context of Global Climate Change." In Guzman, Jose, George Martine, Gordon McGranahan, Daniel

Schensul, and Cecilia Tacoli, eds. 2009. *Population Dynamics and Climate Change.* New York: United Nations Population Fund (UNFPA) & International Institute for Environment and Development (IIED).

Mazur, Laurie, ed. 2009. *A Pivotal Moment: Population, Justice & The Environmental Challenge.* Washington, DC: Island Press.

McCann, Carole R. 2017. *Figuring the Population Bomb: Gender and Demography in the Mid-Twentieth Century.* Seattle, WA: University of Washington Press.

———. 2009. "Malthusian Men and Demographic Transitions: A Case Study of Hegemonic Masculinity in Mid-Twentieth-Century Population Theory." *Frontiers: A Journal of Women Studies* 30(1): 142–71.

McCracken, Mike. 2001. "Prediction Versus Projection-Forecast Versus Possibility." *Weatherzine* 26. www.sciencepolicy.colorado.edu.

McCoy, David, Sudeep Chand, and Devi Sridhar. 2009. "Global Health Funding: How Much, Where it Comes From, and Where it Goes." *Health Policy and Planning* 24(6): 407–17.

McKibben, Bill. 2016. "A World at War." *New Republic*, August 15, 2016. www.newrepublic.com.

———. 2012. "Global Warming's Terrifying New Math." *Rolling Stone*, July 19, 2012. www.rollingstone.com.

Meadows, Donella, Jorgen Randers, and Dennis Meadows. 2004. *Limits to Growth: The 30-Year Update.* White River Junction, VT: Chelsea Green Publishing Company.

Merchant, Carolyn. 2003. "Shades of Darkness: Race and Environmental History." *Environmental History* 8(3): 380–94.

Merchant, Emily R. 2015. *Prediction and Control: Global Population, Population Science, and Population Politics in the Twentieth Century.* Unpublished dissertation filed at University of Michigan.

Mohanty, Chandra Talpade. 1995. "Under Western Eyes: Feminist Scholarship and Colonial Discourses." In Ashcroft, Bill, Garreth Griffiths, and Helen Tiffin, eds. 1995. *The Post-Colonial Studies Reader.* London: Routledge.

Moser, Caroline O.N. 1993. *Gender Planning and Development: Theory, Practice, and Training.* New York: Routledge.

Murphy, Michelle. 2017. *The Economization of Life.* Durham, N.C.: Duke University Press.

Murtaugh, Paul, and Michael Schlax. 2009. "Reproduction and the Carbon Legacies of Individuals." *Global Environmental Change* 19:14–20.

Mutunga, Clive, Eliya Zulu, and Roger-Mark DeSouza. 2012. "Population Dynamics, Climate Change, and Sustainable Development in Africa." Washington, DC: Population Action International & African Institute for Development Policy. www.pai.org.

Mutunga, Clive, and Karen Hardee. 2009. "Population and Reproductive Health in National Adaptation Plans of Action (NAPAs) for Climate Change." In Guzman, Jose, George Martine, Gordon McGranahan, Daniel Schensul, and Cecilia Tacoli, eds. 2009. *Population Dynamics and Climate Change.* New York: United Nations

Population Fund (UNFPA) & International Institute for Environment and Development (IIED).
Nakicenovic, N., et al. 2000. *Special Report on Emissions Scenarios: A Special Report of Working Group III of the Intergovernmental Panel on Climate Change*. Cambridge: Cambridge University Press.
Nelson, Jennifer. 2003. *Women of Color and the Reproductive Rights Movement*. New York: NYU Press.
Nelson, Nicole, Anna Geltzer, and Stephen Hilgartner. 2008. "Introduction: The Anticipatory State: Making Policy-Relevant Knowledge about the Future." *Science and Public Policy* 35(8): 546–50.
Notestein, Frank. 1945. "Population-The Long View." In Schultz, Theodore W., ed. 1945. *Food for the World*. Chicago: University of Chicago Press.
O'Neill, B., M. Dalton, R. Fuchs, L. Jiang, S. Pachauri, and K. Zigova. 2010. "Global Demographic Trends and Future Carbon Emissions." *Proceedings of the National Academy of Sciences* 107(41): 17521–6.
O'Neill, Brian, F. Landis MacKellar, and Wolfgang Lutz. 2001. *Population and Climate Change*. Cambridge: Cambridge University Press.
Osborn, Fairfield. 1953. *The Limits of the Earth*. Boston: Little, Brown.
Pearl, Raymond. 1939. *The Natural History of Population*. New York: Oxford University Press.
———. 1927. "The Growth of Populations." *Quarterly Review of Biology* 2(4): 532–48.
———. 1925. *The Biology of Population Growth*. New York: Alfred A. Knopf, Inc.
———. 1922. "The Population Problem." *Geographical Review* 12(4): 636–45.
Peet, Richard, Paul Robbins, and Michael Watts, eds. 2011. *Global Political Ecology*. London: Routledge.
Petchesky, Rosalind Pollack. 2003. *Global Prescriptions: Gendering Health and Human Rights*. London: Zed Books.
Pielemeier, John. 2005. "Review of Population-Health-Environment Programs Supported by the Packard Foundation and USAID." Private Consultant Report, commissioned by The Packard Foundation and the United States Agency for International Development.
Plautz, Jason. 2014. "The Climate-Change Solution No One Will Talk About." *The Atlantic*, November 1, 2014. www.theatlantic.com.
Porter, Theodore M. 1995. *Trust in Numbers: The Pursuit of Objectivity in Science and Public Life*. Princeton, N.J.: Princeton University Press.
Potts, Malcolm. 2013. "Crisis in the Sahel: Possible Solutions and the Consequences of Inaction." Berkeley: University of California, Berkeley Bixby Center for Population, Health, and Sustainability.
Price, Kimala. 2010. "What is Reproductive Justice?: How Women of Color Activists Are Redefining the Pro-Choice Paradigm." *Meridians* 10(2): 42–65.
Prois, Jessica. "Voluntary Birth Control is a Climate Change Solution Nobody Wants to Talk About." December 4, 2015. www.huffingtonpost.com.

Ralph, Lauren, Sandra McCoy, Karen Shiu, and Nancy Padian. 2015. "Hormonal Contraceptive Use and Women's Risk of HIV Acquisition: A Meta-Analysis of Observational Studies." *The Lancet Infectious Diseases* 15(2): 181–9.

Rao, Mohan, and Sarah Sexton. 2010. "Introduction: Population, Health, and Gender in Neo-liberal Times." In Rao, Mohan and Sarah Sexton. 2010. *Markets and Malthus: Population, Gender, and Health in Neo-liberal Times*. New Delhi: SAGE Publications.

Rathgeber, Eva M. 1990. "WID, WAD, GAD: Trends in Research and Practice." *Journal of Developing Areas* 24(4): 489–502.

Reardon, Jenny. 2004. *Race to the Finish: Identity and Governance in an Age of Genomics*. Princeton, N.J.: Princeton University Press.

Reisch, Michael. 2002. "Defining Social Justice in a Socially Unjust World." *Families in Society: The Journal of Contemporary Social Services* 83(4): 343–54.

Resurreccion, Bernadette. 2013. "Persistent Women and Environment Linkages in Climate Change and Sustainable Development Agendas." *Women's Studies International Forum* 40 (Sept–Oct 2013): 33–43.

Richey, Lisa Ann. 2008. *Population Politics and Development: From the Policies to the Clinics*. New York: Palgrave Macmillan.

Robbins, Paul. 2012. *Political Ecology: A Critical Introduction*. 2nd ed. Sussex, UK: Wiley-Blackwell.

Roberts, Dorothy. 1998. *Killing the Black Body: Race, Reproduction and the Meaning of Liberty*. New York: Vintage Books.

Robertson, Thomas. 2012. *The Malthusian Moment: Global Population Growth and the Birth of American Environmentalism*. New Brunswick, NJ: Rutgers University Press.

Rocheleau, Dianne, Barbara Thomas-Slayter, and Esther Wangari, eds. 1996. *Feminist Political Ecology: Global Issues and Local Experience*. New York: Routledge.

Roe, Emery. 1991. "Development Narratives, or Making the Best of Blueprint Development." *World Development* 19(4): 287–300.

Roosevelt, Theodore. 1908. "Opening Address by the President." Proceedings of a Conference of Governors. Washington, DC: Government Printing Office. www.memory.loc.gov.

———. 1905. "On American Motherhood." The World's Famous Orations. America: III. (1861-1905). Bartleby.com. www.bartleby.com.

Sasser, Jade. 2016. "Population, Climate Change, and the Embodiment of Environmental Crisis." In Godfrey, Phoebe, and Denise Torres, eds. 2016. *Systemic Crises of Global Climate Change: Intersections of Race, Class, and Gender*. New York: Routledge.

———. 2014a. "From Darkness Into Light: Race, Population, and Environmental Advocacy." *Antipode* 46(5): 1240–57.

———. 2014b. "The Wave of the Future? Youth Advocacy at the Nexus of Population and Climate Change." *Geographical Journal* 180(2): 102–10.

———. 2009. "Environmental Organizations and Population Programs: A Preliminary Analysis." Berkeley: Venture Strategies for Health & Development.

Satterthwaite, David. 2009. "The Implications of Population Growth and Urbanization for Climate Change." *Environment and Urbanization* 21(2): 545–67.

Sayre, Nathan. 2008. "The Genesis, History, and Limits of Carrying Capacity." *Annals of the Association of American Geographers* 98(1): 120–34.

Seager, Joni. 2009. "Death by Degrees: Taking a Feminist Hard Look at the 2° Climate Policy." *Kvinder, Køn og Foraksning* 3–4: 11–21.

Sen, Gita. 1994. "Women, Poverty and Population: Issues for the Concerned Environmentalist." In Harcourt, Wendy, ed. 1994. *Feminist Perspectives on Sustainable Development.* London: Zed Books.

Sen, Gita, and Karen Grown. 1987. *Development, Crises, and Alternative Visions: Third World Women's Perspectives.* New York: Monthly Review Press.

Sen, Gita, and Srilatha Batliwala. 2000. "Empowering Women for Reproductive Rights." In Presser, Harriet, and Gita Sen. 2000. *Women's Empowerment and Demographic Processes: Moving Beyond Cairo.* New York: Oxford University Press.

Sharpless, John. 1997. "Population Science, Private Foundations, and Development Aid: The Transformation of Demographic Knowledge in the United States, 1945–1965." In Cooper, Frederick, and Randall Packard, eds. 1997. *International Development and the Social Sciences: Essays on the History and Politics of Knowledge.* Berkeley: University of California Press.

Silliman, Jael, and Ynestra King. 1999. *Dangerous Intersections: Feminist Perspectives on Population, Environment, and Development.* Cambridge, MA: South End Press.

Silliman, Jael, Marlene Gerber Fried, Loretta Ross, and Elena Gutierrez. 2004. *Undivided Rights: Women of Color Organize for Reproductive Justice.* Cambridge, MA: South End Press.

Slater, Jack. 1973. "Sterilization: Newest Threat to the Poor." *Ebony Magazine* 28: 150–6.

Smith, Andrea. 2015 [2005]. *Conquest: Sexual Violence and American Indian Genocide.* Durham, NC: Duke University Press.

Solinger, Rickie. 2005. *Pregnancy and Power: A Short History of Reproductive Politics in America.* New York: NYU Press.

Speidel, J. Joseph, Stephen Sinding, Duff Gillespie, Elizabeth Maguire, and Margaret Neuse. 2008. "Making the Case for U.S. International Family Planning Assistance." Baltimore: Johns Hopkins University Gates Institute. www.jhsph.edu.

Standing, Guy. 2001. *Globalisation: The Eight Crises of Social Protection.* Geneva: International Labour Organization.

Stern, Alexandra Minna. 2005. *Eugenic Nation: Faults & Frontiers of Better Breeding in Modern America.* Berkeley: University of California Press.

Stirling, Andy. 2007. "Deliberate Futures: Precaution and Progress in Social Choice of Sustainable Technology." *Sustainable Development* 15: 286–95.

Sullivan, Rachel. 2007. *Leveraging the Global Agenda for Progress: Population Policies and Non-Governmental Organizations in Sub-Saharan Africa.* Unpublished dissertation filed at University of California, Berkeley.

Summers, Lawrence. 1992. "Investing in *All* the People." Policy Research Working Paper. Washington, DC: The World Bank.

Szreter, Simon. 1993. "The Idea of Demographic Transition and the Study of Fertility Change: A Critical Intellectual History." *Population and Development Review* 19(4): 659–701.

Tabutin, Dominique, and Bruno Schoumaker. 2004. "The Demography of Sub-Saharan Africa from the 1950s to the 2000s: A Survey of Changes and a Statistical Assessment." *Population* (English Edition) 59(3): 457–555.

Takeshita, Chikako. 2012. *The Global Biopolitics of the IUD*. Cambridge, MA: MIT Press.

Tapia Granados, Jose, Edward Ionides, and Oscar Carpintero. 2012. "Climate Change and the World Economy: Short-Run Determinants of Atmospheric CO2." *Environmental Science & Policy* (21): 50–62.

Turshen, Meredith. 2007. *Women's Health Movements: A Global Force for Change*. New York: Palgrave Macmillan.

United Nations. 2015. United Nations Social & Economic Affairs. World Population Prospects. 2015 Revision. www.esa.un.org.

———. 2004 [1994]. Programme of Action: Adopted at the International Conference on Population and Development, Cairo, 5–13 September 1994. New York: UNFPA.

———. 2004. UNRISD Conference on Social Knowledge and International Policy Making: Exploring the Linkages, 20–21 April 2004. Geneva, Switzerland.

———. 1992. The Rio Declaration on Environment and Development in *Report of the United Nations Conference on Environment and Development*. Rio de Janiero: United Nations.

Unnamed author. 2016. "COP16 Policy Statement—Global Youth Support Sexual and Reproductive Health and Rights (SRHR) for a Just and Sustainable World." www.advocatesforyouth.org.

Van Vorst, Mrs. John, and Marie Van Vorst. 2010 [1903]. *The Woman Who Toils: Being the Experiences of Two Ladies as Factory Girls*. Carlisle, MA: Applewood Books.

Various authors. 1992. *Treaty on Population, Environment and Development*. Produced at the United Nations Conference on Environment and Development (UNCED), Rio, Brazil. www.stakeholderforum.org.

Verilli, Adrienne, and Julia Piscitelli. 2008. "Research Findings Report-New Population Challenge." Washington, DC: Spitfire Strategies.

Vogt, William. 1948. *Road to Survival*. New York: William Sloane Associates, Inc.

Warren, Julianne Lutz. 2016. *Aldo Leopold's Legacy*. Washington, DC: Island Press.

Watts, Michael J. 1983. *Silent Violence: Food, Famine, and Peasantry in Northern Nigeria*. Berkeley: University of California Press.

Whitty, Julia. 2010. "The Last Taboo." *Mother Jones*. (May/June 2010): 24–43.

Wilson, Edward W., and Keiki Kehoe. 2000. "Environmental Organizations and International Population Assistance." Special Report Submitted to the Summit, Hewlett & Packard Foundations.

Wilson, Kalpana. 2015. "Towards a Radical Re-appropriation: Gender, Development, and Neoliberal Feminism." *Development and Change* 46(4): 803–32.

Win, Everjoice. 2007. "Not Very Poor, Powerless, or Pregnant: The African Woman Forgotten by Development." In Cornwall, Andrea, Elizabeth Harrison, and Ann Whitehead, eds. 2007. *Feminisms in Development: Contradictions, Contestations & Challenges*. London: Zed Books.

Wire, Thomas. 2009. "Fewer Emitters, Lower Emissions, Less Cost: Reducing Future Carbon Emissions by Investing in Family Planning: A Cost/Benefit Analysis." Report produced for Optimum Population Trust. London.

Wohlforth, Charles. 2010. "Conservation and Eugenics: The Environmental Movement's Dirty Little Secret." *Orion* (July/Aug 2010): 22–28.

World Bank Open Data, 2015. www.data.worldbank.org.

World Health Organization. 2017. "Hormonal Contraceptive Eligibility for Women at High Risk of HIV: Guidance Statement." Department of Reproductive Health and Research. Geneva, Switzerland: World Health Organization. www.apps.who.int.

Zelman, Joanna. "World Population Day: A Connection Between Global Warming and Overpopulation?" *Huffington Post*, July 11, 2011. www.huffingtonpost.com.

ZPG Reporter. 1974. "ZPG Goals." *Zero Population Growth National Reporter* 6(3): 2.

INDEX

ABOUT THE AUTHOR

Jade S. Sasser is Assistant Professor in the Department of Gender and Sexuality Studies at the University of California, Riverside.